**Solving Problems in Applied Thermodynamics
and Energy Conversion**

Other titles in the Series

Solving Problems in Vibrations, *J.S. Anderson and M. Bratos-Anderson*

Solving Problems in Structures Volume 1, *P.C.L. Croxton and L.H. Martin*

Solving Problems in Fluid Mechanics Volume 1, *J.F. Douglas*

Solving Problems in Fluid Mechanics Volume 2, *J.F. Douglas*

Solving Problems in Soil Mechanics, *B.H.C. Sutton*

Solving Problems in Solid Mechanics Volume 1, *S.A. Urry and P.J. Turner*

Solving Problems in Solid Mechanics Volume 2, *S.A. Urry and P.J. Turner*

Solving Problems in Applied Thermodynamics and Energy Conversion

G.J. Sharpe BSc(Eng), BSc, MPhil, CEng, MIMechE, FInstE

Copublished in the United States with
John Wiley & Sons, Inc., New York

Longman Scientific & Technical,
Longman Group UK Limited,
Longman House, Burnt Mill, Harlow,
Essex CM20 2JE, England
and Associated Companies throughout the world.

Copublished in the United States with
John Wiley & Sons Inc., 605 Third Avenue, New York,
NY 10158

© G.J. Sharpe, 1987

All rights reserved; no part of this publication may be reproduced, stored in a retrieval system, or transmitted in any form or by an means, electronic, mechanical, photocopying, recording, or otherwise, without the prior written permission of the Publishers.

First published 1987

British Library Cataloguing in Publication Data
Sharpe, G.J.
   Solving problems in applied thermodynamics
   and energy conversion. − (Solving problems)
   1. Power (Mechanics) − Problems, exercises
   etc.
   I. Title
621.4   TJ163.9

ISBN 0-582-28647-6

Library of Congress Cataloging-in-Publication Data
Sharpe, G.J. (George Joseph)
   Solving problems in applied thermodynamics and
energy conversion.

   (Solving problems in)
   Bibliography: p.
   Includes index.
   1. Thermodynamics.   2. Power (Mechanics)   I. Title.
II. Series.
TJ265.S487 1987   621.402'1   86-33767
ISBN 0-470-20707-8 (USA only)

Set in Times 10/12 pt

Produced by Longman Singapore Publishes (Pte) Ltd.
Printed in Singapore.

Contents

Preface vii

Units and conversion factors ix

List of symbols xi

1. Combustion 1
2. Nozzles: rocket propulsion 18
3. Vapour cycles 33
4. Internal combustion engines 51
5. Gas turbines 72
6. Turbomachinery: reciprocating machinery 91
7. Heat transfer 124
8. Psychrometry: refrigeration, heat pumps 155
9. Alternative (renewable) sources of energy 172
10. Waste-heat recovery; total energy: combined heat and power; energy economics 191

Further reading 214

Index 215

Preface

This volume contains the material required for BSc and Higher Diploma courses in Mechanical Engineering, in which a knowledge of applied thermodynamics is required. The emphasis is placed on the energy-conversion aspects. Consequently the material can be of assistance to other courses in which energy conversion and utilization is studied, such as chemical engineering, building-services engineering, plant engineering, environmental science.

The treatment adopted is the same as that used in the *Solution of Problems* series. Each chapter comprises a summary of the basic theory, a selection of worked examples to illustrate the theoretical work, and several examples (with answers) for the reader to attempt. It should be noted that the theory is in **summary** only, and it is assumed that the reader has some knowledge of thermodynamic theory. The volume is therefore intended to complement a text book.

The majority of the problems have been devised by the author for use in examinations and tutorial work. Some have been taken from examination papers set by Newcastle upon Tyne Polytechnic, Brunel University, the University of Surrey, the National University of Singapore, and the former Council of Engineering Institutions. I am indebted to these bodies for permission to use these questions.

Inevitably a few errors may have escaped the scrutiny of the author, publisher and printer. I should be pleased to receive any corrections or constructive criticism.

G.J. Sharpe

Units and conversion factors

The SI is used in this volume, but in the field of energy conversion and utilization non-SI units are still in use. The SI units and conversion factors are therefore listed for the convenience of the reader.

Quantity	*Unit*	*Symbol or abbreviation*
length	metre	m
mass	kilogramme	kg
time	second	s
electric current	ampere	A
temperature	degree Kelvin	K
density		kg/m^3
momentum		kg m/s
force	newton	N
pressure	pascal	N/m^2
work, energy, heat	joule	J
power	watt	W
specific entropy		J/kg K

Common multiples

10^{12}	tera	T	10^{-1}	deci	d
10^{9}	giga	G	10^{-2}	centi	c
10^{6}	mega	M	10^{-3}	milli	m
10^{3}	kilo	k	10^{-6}	micro	μ

Conversion factors

Length	: 1 mile = 1.609 km
	: 1 inch = 25.4 mm
Capacity	: 1 gallon = 4.545 dm^3 (litres)
	: 1 US gallon = 3.785 dm^3
Mass	: 1 ton = 1016 kg = 1.016 tonne
	: 1 lb = 0.4536 kg
Specific volume	: 1 ft^3/lb = 62.43 dm^3/kg
Fuel consumption	: 1 mile/gal = 0.354 km/l (dm^3)
Density	: 1 lb/ft^3 = 16.019 kg/m^3
Force	: 1 lbf = 4.448 N
Torque	: 1 lbf ft = 1.3558 Nm
Pressure	: 1 lbf/in^2 = 6895 N/m^2
	: 1 in H$_2$O = 249 N/m^2
	: 1 bar = 10^5 N/m^2
	: 1 atm = 1.013 bar = 101.3 kN/m^2
	: 1 mm Hg (torr) = 133.3 N/m^2
Viscosity	: 1 poise = 0.1 N s/m^2
	: 1 lb/ft s = 1.488 kg/m s = 1.488 poise
	: 1 stoke = 10^{-4} m^2/s
	: 1 ft^2/s = 0.093 m^2/s
Energy	: 1 therm = 105.5 MJ
	: 1 Btu = 1.055 kJ
	: 1 kWh = 3.6 MJ
Power	: 1 hp = 0.7457 kW
Specific energy	: 1 Btu/lb = 2.326 kJ/kg
	: 1 Btu/ft^3 = 37.26 kJ/m^3
Thermal capacity	: 1 Btu/lb deg F = 4.187 kJ/kg K
Conductance	: 1 Btu/ft^{2}h = 3.1546 W/m^2
	: 1 Btu/ft^{2}h deg F = 5.6783 W/m^2 K

List of symbols

The symbols listed are those which appear throughout the book. Other symbols are stated in the appropriate chapters.

a	acoustic velocity
A	air:fuel ratio
A_0	stoichiometric air:fuel ratio
ASE	air standard efficiency
Bi	Biot number
c	effective exit velocity
C	cost
C_p	specific heat at constant pressure; pressure coefficient
C_F	thrust coefficient
CR	compression ratio
C_v	specific heat at constant volume; velocity coefficient
COP	coefficient of performance
D	diameter
e	emissivity; effectiveness
E	Euler head; emission; energy consumption
F	thrust; geometrical factor
Fo	Fourier number
Gr	Grashof number
h	enthalpy
h_c	convection coefficient
H	head
I	incident radiation
I_s	specific impulse
J	radiosity
K	thermal conductivity coefficient
K_p	equilibrium constant
L	stroke; beam length
m	mass
M	Mach number
MEP	mean effective pressure
N	speed
N_s	specific speed
Nu	Nusselt number
p	pressure
P	power
P_s	specific power

Pr	Prandtl number
Q	heat; flow rate
r_p	pressure ratio
R	gas constant; thermal resistance
Re	Reynolds number
s	entropy
S	energy consumption rate
SFC	specific fuel consumption
SSC	specific steam consumption
T	temperature
u	internal energy; velocity
U	overall coefficient of heat transfer
v	volume; velocity
W	work
W_s	shaft work

Greek letters

α	degree of dissociation; absorptivity
β	coefficient of thermal expansion
γ	specific heat ratio
Δ	change in
η	efficiency
η_c	Carnot cycle; compressor
η_D	diffuser
η_f	fin
η_h	hydraulic
η_N	nozzle
η_p	propulsive
η_R	Rankine cycle
η_t	turbine
η_V	volumetric
θ	blade angle
λ	work done factor
μ	absolute (dynamic) viscosity
ν	kinematic viscosity
ρ	density
σ	Stefan constant
τ	transmittivity
ϕ	relative humidity
ω	specific humidity

Subscripts

Two subscripts used throughout are:

s denoting constant entropy conditions
0 denoting stagnation conditions

1

Combustion

Combustion is one of the important energy-conversion processes, in which the energy in a fuel is released by an exothermic chemical reaction. This thermal energy can then be converted into mechanical–kinetic–thermal energy in many systems, such as steam plant, I.C. engines, gas turbines, rocket propulsion.

The main combustible elements (except for special conversions such as rockets) are carbon and hydrogen, and there are also incombustibles such as water and mineral matter. The latter forms ash when the fuel is burned. The many types of fuels are as grouped below:

Fossil
- solid – peat, lignite, coal
- liquid – petroleum, shale oil
- gaseous – natural gas

Secondary
- wood, refuse, coke, coal-tar fuel
- mineral and vegetable oils, petroleum products
- coal gas, producer gas, substitute natural gas

The chemical reaction requires a source of oxygen for combustion to take place, and this is usually air. In some special plants the oxidant is liquid oxygen.

Calorific value

The calorific value (CV) of a fuel is the heat released by the fuel when completely burnt, and may be determined at constant volume (bomb calorimeter) or constant pressure. The hydrogen in the fuel will burn to form steam, and the higher or gross calorific value includes the latent heat of the steam. The lower or net calorific value does not include this latent heat.

The net CV can be calculated from the gross CV, which is the value determined by the use of a calorimeter, by the standard equation:

net CV = gross CV − 41.4 MJ/kmol H_2O

Air : fuel ratio

The combustion equation for a fuel burned in air can be written in the form

COMBUSTION 1

$$1.0[\text{fuel}] + 0.21A\,[O_2] + 0.79A\,[N_2] \rightarrow$$
$$a\,[CO_2] + b\,[CO] + c\,[O_2] + d\,[N_2] + e\,[H_2O] + f\,[SO_2]$$

where A = kmol air supplied/kmol fuel
a, b, c, d, e, f = kmol product/kmol fuel

The following basic data should be noted

1 kmol = molecular wt (kg) = 22.41 m³ at 1 atm, 0 °C
= 22.41 nm³

Air composition is

by mass   23.3 % O_2, 76.7 % N_2
by volume   21.0 % O_2, 79.0 % N_2

Molecular weights are

Carbon	C = 12	Nitrogen	N_2 = 28
Hydrogen	H_2 = 2	Sulphur	S = 32
Oxygen	O_2 = 32	Air	= 29

The stoichiometric air, A_0, is the minimum theoretical amount of air required to burn a fuel completely. In this case $b = c = 0$, and the values of a, d, e, f, A_0 can be determined from a balance of the elements C, H_2, O_2, N_2, S.

In practice the air supplied is greater than this to obtain complete combustion, due to problems involved in complete mixing of the air and fuel, time taken to burn the fuel, and non-uniform temperature distributions. In this case $A > A_0$, and there is likely to be unburnt carbon (appearing as CO) and excess or unused oxygen in the combustion products. The determination of the air supplied requires a knowledge of the analysis of the products. This would normally be the percentage of CO_2, CO and sometimes O_2 in the products, measured by volume on a *dry* products basis.

Thus

$$\% \; CO_2 = \frac{100a}{a+b+c+d+f}$$

The excess air is measured as a percentage of the stoichiometric air, i.e.

$$\% \text{ excess air} = 100\,(A - A_0)/A_0$$

Dissociation At high temperatures of combustion dissociation of the products becomes significant. For example, consider the combustion of carbon monoxide:

$$CO + \tfrac{1}{2}O_2 \rightleftharpoons CO_2$$

The reaction proceeds in both directions, and a fraction, α, of the CO_2 produced is dissociated into CO and O_2:

$$\alpha CO_2 \rightarrow \alpha(CO + \tfrac{1}{2}O_2)$$

In the equilibrium condition the products consist of

$$(1 - \alpha)CO_2 + \alpha CO + \tfrac{1}{2}\alpha O_2$$

The degree of dissociation, α, can be determined from the total pressure of the products mixture P, and the equilibrium constant K_p (which is a function of the temperature). In the reaction quoted

$$K_p = P_{CO_2}/P_{CO}\sqrt{P_{O_2}}$$

where p_x = the partial pressure of component x.

Hence: $$p_{CO_2} = \frac{1 - \alpha}{1 - \alpha + \alpha + \tfrac{1}{2}\alpha}P = \frac{1 - \alpha}{1 + \tfrac{1}{2}\alpha}P$$

$$p_{CO} = \frac{\alpha P}{1 + \tfrac{1}{2}\alpha}, \text{ and } p_{O_2} = \frac{\tfrac{1}{2}\alpha P}{1 + \tfrac{1}{2}\alpha}$$

$$\therefore \quad K_p = \frac{1 - \alpha}{\alpha}\sqrt{\left(\frac{\alpha + 2}{\alpha P}\right)}$$

In general terms for a reaction given by

$$a \cdot A + b \cdot B = c \cdot C + d \cdot D$$

the equilibrium constant K_p

$$= \frac{p_C^c p_D^d}{p_A^a p_B^b} \cdot P^{(c+d)-(a+b)}$$

The effect of dissociation is to reduce the flame or products temperature, since heat is absorbed (endothermic process) in the dissociation.

Adiabatic flame temperature

In an adiabatic combustion process the temperature of the products T_f is greater than that of the reactants T_r due to the heat released by the reaction. The steady flow energy gives

$$m_p h_p = m_r h_r - Q_p$$

where Q_p = heat of reaction at constant pressure. It should be noted that Q_p is negative for an exothermic reaction.

In the case of combustion at constant volume

$$m_p u_p = m_r u_r - Q_v$$

where u = internal energy, Q_v = heat of reaction at constant volume.
Also $Q_p - Q_v = (n_p - n_r)R_0 T$

where n = number of kmol of the products and reactants. The heat of reaction is normally taken at 25 °C.

The determination of the flame temperature involves an iterative solution since the enthalpy $h = C_p T$, and C_p is a function of the temperature.

Combustion formulae

An approximate formula for the excess air is

$$\% \text{ excess air} = 100\left[\frac{(CO_2)_0}{(CO_2) + (CO)} - 1\right]$$

or $= 100 \left[\dfrac{(O_2) - \frac{1}{2}(CO)}{21 - (O_2)} \right]$

where $(CO_2)_0 = \%\ CO_2$ in the stoichiometric dry products
$CO_2, CO, O_2 = \%$ in the actual products

A formula that can be used to check the accuracy of the flue gas analysis is

$$(CO_2)_0 = \dfrac{21\,[(CO_2) + (CO)]}{21 - (O_2) + 0.395\,(CO)}$$

1.1 Excess air, solid fuel

> The analysis, by mass (gravimetric) of an anthracite fuel is 90 % carbon, 3 % hydrogen, 2 % oxygen, 1 % nitrogen. Calculate the stoichiometric air:fuel ratio (kg/kg).
>
> Given that the analysis of the dry combustion products, by volume, is 16.2 % CO_2, 3.5 % O_2, 80.3 % N_2, calculate the air:fuel ratio, and % excess air used.
>
> Also calculate the % CO_2 in the dry stoichiometric products, and hence the ratio % $(CO_2)_0:(CO_2)$. Comment on the value obtained.

Solution Consider 1 kg of fuel. The combustion equation, with all quantities expressed in **kmol**, is

$$\dfrac{0.90}{12}[C] + \dfrac{0.03}{2}[H_2] + \dfrac{0.02}{32}[O_2] + \dfrac{0.01}{28}[N_2]$$
$$+ 0.21 A_0[O_2] + 0.79 A_0[N_2] \to a[CO_2] + d[N_2] + e[H_2O]$$

for stoichiometric combustion (no CO or O_2 in the products).

A carbon balance gives $\quad a = 0.90/12 = 0.0750$
hydrogen balance gives $\quad e = 0.03/2 = 0.0150$
oxygen balance gives $\quad a + \frac{1}{2}e = 0.02/32 + 0.21 A_0$
nitrogen balance gives $\quad d = 0.01/28 + 0.79 A_0$

Solving gives: $A_0 = (0.0750 + 0.0075 - 0.0006)/0.21$
$\qquad\qquad\quad = 0.39$ kmol
$\qquad\qquad\quad = 0.39(29) = $ **11.31 kg/kg**

The combustion equation, non-stoichiometric, is

$$\dfrac{0.90}{12}[C] + \dfrac{0.03}{2}[H_2] + \dfrac{0.02}{32}[O_2] + \dfrac{0.01}{28}[N_2]$$
$$+ 0.21 A[O_2] + 0.79 A[N_2]$$
$$\to a[CO_2] + b[CO] + c[O_2] + d[N_2] + e[H_2O]$$

In this problem there is no CO in the products, therefore $b = 0$. The element balances are:

carbon $a = 0.075$
hydrogen $e = 0.015$
oxygen $a + c + \tfrac{1}{2}e = 0.0006 + 0.21A$
nitrogen $d = 0.0004 + 0.79A$

Also % CO_2 (dry products) $= 16.2 = \dfrac{100a}{a+c+d}$

% O_2 (dry products) $= 3.5 = \dfrac{100c}{a+c+d}$

Hence $\dfrac{a}{c} = \dfrac{16.2}{3.5}$

$\therefore c = \dfrac{3.5}{16.2}(0.075) = 0.162$

and $0.21A = 0.075 + 0.0162 + \tfrac{1}{2}(0.015) - 0.0006$

$\therefore A = 0.4671$ kmol $= \mathbf{13.55}$ **kg/kg**

% excess air $= 100(A - A_0)/A_0$
$= 100(13.55 - 11.31)/11.31 = \mathbf{19.8}$

In the dry stoichiometric products, the % CO_2 is

$$(CO_2)_0 = \dfrac{100A}{a+d} = \dfrac{100(0.075)}{0.075 + 0.01/28 + 0.79(0.390)} = 19.56$$

$(CO_2)_0/CO_2 = 19.56/16.2 = \mathbf{1.207}$

The value obtained for this ratio of the CO_2 is 1.207, and an approximate formula for the % excess air is

% excess air $= 100[(CO_2)_0/CO - 1]$

In this case a value of $100(1.207) - 1 = 20.7$ % would be obtained, which compares favourably with the calculated value of 19.8 %. This formula is generally a good approximation, although the discrepancy is greater with gaseous fuels.

1.2 Excess air, gaseous fuel, dew points

> The volumetric analysis of a gaseous fuel is 25 % CH_4, 45 % H_2, 3 % CO_2, 10 % CO, 1 % O_2, 1 % C_2H_4, 15 % N_2. Calculate the stoichiometric air:fuel ratio, m^3/m^3.
>
> The dry combustion products contain, by volume, 7.11 % CO_2 and 0.31 % CO. Calculate the air:fuel ratio, % excess air used, and % O_2 in the dry products. Also determine the theoretical dew point.

Solution In the case of gaseous fuels, the analysis is by volume and therefore the same in kmol. The stoichiometric combustion equation/kmol fuel is

$$0.25[CH_4] + 0.45[H_2] + 0.03[CO_2] + 0.10[CO]$$
$$+ 0.01[O_2] + 0.01[C_2H_4] + 0.15[N_2] + 0.21A_0[O_2]$$
$$+ 0.79A_0[N_2] \rightarrow a[CO_2] + d[N_2] + e[H_2O]$$

Balancing the elements:

C $0.25 + 0.03 + 0.10 + 2(0.01)$ $= a$
H_2 $2(0.25) + 0.45 + 2(0.01)$ $= e$
O_2 $0.03 + \frac{1}{2}(0.10) + 0.01 + 0.21A_0 = a + \frac{1}{2}e$

Therefore, $a = 0.400$, $e = 0.970$, $A_0 = 3.786$ kmol
Therefore, stoichiometric air:fuel ratio = **3.786 m³/m³**

With excess air the combustion equation becomes

$$0.25[CH_4] + 0.45[H_2] + 0.03[CO_2] + 0.10[CO]$$
$$+ 0.01[O_2] + 0.01[C_2H_4] + 0.15[N_2] + 0.21A[O_2]$$
$$+ 0.79A[N_2] \rightarrow a[CO_2] + b[CO] + c[O_2] + d[N_2] + e[H_2O]$$

Balancing the elements

C $0.25 + 0.03 + 0.10 + 2(0.01)$ $= a + b = 0.400$
H_2 $2(0.25) + 0.45 + 2(0.01)$ $= e = 0.970$
O_2 $0.03 + \frac{1}{2}(0.10) + 0.01 + 0.21A = a + \frac{1}{2}b + c + \frac{1}{2}e$
N_2 $0.15 + 0.79A$ $= d$

Now % $CO_2 = 7.11 = \dfrac{100a}{a+b+c+d}$

% $CO = 0.31 = \dfrac{100b}{a+b+c+d}$

Therefore $\dfrac{b}{a} = \dfrac{0.31}{7.11} = 0.0436$ and $a + b = 0.400$

Therefore $a = 0.3833$, $b = 0.0167$.

Subst. $0.21A = a + \frac{1}{2}b + c + \frac{1}{2}e - 0.09$
$0.79A = d - 0.15$

But $a + b + c + d = \dfrac{100a}{7.11} = 5.3910$, therefore $c + d = 4.9910$

Adding the two previous equations

$A = a + \frac{1}{2}b + c + \frac{1}{2}e - 0.09 + d - 0.15 =$ **5.628 kmol**

% excess air $= 100(A - A_0)/A_0 = 100(5.628 - 3.786)/3.786$
$=$ **48.7**

% O_2 in dry products $= \dfrac{100c}{a+b+c+d} = \dfrac{100(0.3949)}{0.400 + 4.9910} = 7.33$

since $d = 0.15 + 0.79A = 4.5961$, therefore
$c = 4.9910 - d = 0.3949$

The theoretical dew point is the temperature at which the steam in the combustion products just begins to condense to liquid. It is therefore equal to the saturation temperature corresponding to the *partial* pressure of the H_2O, which in this problem is

$$\frac{e}{a+b+c+d+e} = \frac{0.970}{5.3910 + 0.970} \times 1.013 = 0.154 \text{ bar}$$

The total pressure of the products is taken as 1 atm = 1.013 bar. From steam tables dew point = **55 °C**.

1.3 Analysis of combustion products

> Petrol, gravimetric analysis, 85 % C, 15 % H_2 is burned in an engine with an air:fuel ratio of 13.6 kg/kg. Assuming that all the hydrogen burns to water, and that there is no free oxygen in the combustion products, determine the volumetric analysis of the dry products.
>
> The engine runs on a four-stroke cycle, and has four cylinders. The air is drawn into the cylinder at 1.0 bar, 20 °C. The cylinder bore = 100 mm, stroke = 120 mm. Speed = 3000 rev/min. Determine the petrol consumed, kg/s.

Solution Consider 1 kg petrol. The combustion equation with all quantities in kmol is

$$\frac{0.85}{12}[C] + \frac{0.15}{2}[H_2] + 0.21\left(\frac{13.6}{29}\right)[O_2]$$

$$+ 0.79\left(\frac{13.6}{29}\right)[N_2] \rightarrow a[CO_2] + b[CO] + d[N_2] + e[H_2O]$$

Balancing the elements

C $0.85/12 = a + b = 0.0708$
H_2 $0.15/2 = e = 0.0750$
O_2 $0.0985 = a + \tfrac{1}{2}b + \tfrac{1}{2}e$
N_2 $0.3705 = d$

Therefore, $a + \tfrac{1}{2}b = 0.0985 - \tfrac{1}{2}e = 0.0610$
$a = 0.0512, b = 0.0196$

The dry products are

CO_2	0.0512 kmol	11.60 %
CO	0.0196 kmol	4.44 %
N_2	0.3705 kmol	83.96 %
	0.4413	

Speed = 3000 rev/min = 6000 strokes/min = 100 strokes/s.
Therefore, suction strokes (4-stroke cycle) = 50 strokes/s.
Therefore, swept volume (assumed a volumetric efficiency of 100 %)

$$= 4 \times \frac{\pi}{4}(0.1)^2 \times 0.12 \times 50 = 0.1885 \text{ m}^3/\text{s}$$

Air:fuel ratio = 13.6 kg/kg, therefore the volume of suction air/kg petrol

$$= \frac{13.6 \times 287 \times 293}{1.0 \times 10^5} = 11.436 \text{ m}^3$$

Therefore, fuel consumption = 0.1885/11.436 = **0.0165 kg/s**.

1.4 Dissociation

> Carbon monoxide is burnt with oxygen as a stoichiometric mixture, in a rigid vessel. The initial conditions of the reactants are 1 atm, 330 K.
>
> Calculate the analysis of the combustion products and the pressure at 2500 K.
>
> The equilibrium constant $= \dfrac{p_{CO_2}}{p_{CO}\sqrt{p_{O_2}}} = 27.039 \text{ atm}^{-1/2}$

Solution The initial conditions are $p_1 = 1$ atm, $T_1 = 330$ K and the kmol reactants = 1.5: this follows from the stoichiometric equation

$$CO + \tfrac{1}{2}O_2 \rightarrow CO_2$$

Allowing for dissociation the combustion equation can be written as $CO + \tfrac{1}{2}O_2 \rightarrow (1-\alpha)CO_2 + \alpha CO + \tfrac{1}{2}\alpha O_2$. The kmol products $N_2 = (1-\alpha) + \alpha + \tfrac{1}{2}\alpha = 1 + \tfrac{1}{2}\alpha$. Using the state equation $pv = NR_0 T$, where R_0 = universal gas constant (kJ/kmol K)

$$p_1 v_1 = N_1 R_0 T_1 \quad \text{and} \quad p_2 v_2 = N_2 R_0 T_2$$

Therefore, $\dfrac{p_2}{p_1} = \dfrac{N_2 T_2}{N_1 T_1}$ since the volume is constant.

Therefore, $p_2 = \dfrac{(1 + \tfrac{1}{2}\alpha)(2500)}{1.5(330)} = 5.0505(1 + \tfrac{1}{2}\alpha)$ atm.

The degree of dissociation, α, is determined from the equilibrium constant

$$K_p = 27.039 = \frac{(1-\alpha)p_2}{1 + \tfrac{1}{2}\alpha} \cdot \frac{1 + \tfrac{1}{2}\alpha}{\alpha p_2} \sqrt{\left(\frac{1 + \tfrac{1}{2}\alpha}{\tfrac{1}{2}\alpha p_2}\right)}$$

$$= \frac{1 - \alpha}{\alpha} \sqrt{\left(\frac{1 + \tfrac{1}{2}\alpha}{\tfrac{1}{2}\alpha p_2}\right)}$$

Subst. for p gives $27.039 = \dfrac{1-\alpha}{\alpha} \sqrt{\left(\dfrac{1 + \tfrac{1}{2}\alpha}{\tfrac{1}{2}}\right)} \times \dfrac{1}{5.0505(1 + \tfrac{1}{2}\alpha)}$

$$= \frac{1-\alpha}{\alpha} \sqrt{\left(\frac{0.396}{\alpha}\right)}$$

Solving, by trial and error, gives $\alpha = $ **0.077**.

The products are

CO_2	$1 - \alpha =$	0.923 kmol	88.92 %
CO	$\alpha =$	0.077 kmol	7.42 %
O_2	$\tfrac{1}{2}\alpha =$	0.038 kmol	3.66 %
		1.038	

Also $N_2 = 1.038$ kmol, therefore $p_2 = 5.0505(1.038) =$ **5.242 atm**.

1.5 Adiabatic flame temperature

> A stoichiometric mixture of methane and oxygen, at 25 °C, is burnt at a constant pressure of 1 atm. The heat of reaction is -802.3 MJ/kmol. Estimate the adiabatic flame temperature.
> Outline how the effect of dissociation can be allowed for, deriving the equations required to obtain a value of the flame temperature. Do not attempt to solve the equations.

Solution The stoichiometric combustion equation is

$$CH_4 + 2O_2 \rightarrow 2CO_2 + 2H_2O + 802.3 \text{ MJ}$$

The adiabatic flame temperature T_f is determined from the energy balance

$$802\,300 = (\bar{C}_{pCO_2} + 2 \cdot \bar{C}_{pH_2O})(T_f - 298) \text{ kJ}$$

where $\bar{C}_p$ = mean specific heat over the temperature range 298 to T_f (kJ/kmol). The values are obtained from thermodynamic tables. The solution requires an iterative calculation, since C_p varies with temperature. Hence a value of T_f is assumed, the mean specific heats obtained, and the values substituted in the equation.

T_f (K)	4000	5800	5900
$\bar{C}_{pCO_2}$ (kJ/kmol)	50.226	50.99	51.04
$\bar{C}_{pH_2O}$ (kJ/kmol)	45.729	46.83	46.88
$\Sigma \bar{C}_p (T_f - 298)$	524 510	795 810	811 120

Therefore, **$5800 < T_f < 5900$ K**.

If dissociation is taken into account, the reactions involved are $CO_2 \rightleftharpoons CO + \tfrac{1}{2}O_2$ and $H_2O \rightleftharpoons H_2 + \tfrac{1}{2}O_2$. The overall combustion equation now becomes

$$CH_4 + 2O_2 \rightarrow aCO + bCO_2 + cO_2 + dH_2 + eH_2O$$

There are now five unknowns, and they can be determined from five equations — three derived from balances of the elements, and two from the equilibrium constants.

C balance	$a + b = 1$
H_2 balance	$d + e = 2$
O_2 balance	$\tfrac{1}{2}a + b + c + \tfrac{1}{2}e = 2$

$$K_1 = \frac{p_{CO}\sqrt{(p_{O_2})}}{p_{CO_2}} = \frac{\frac{aP}{N}\sqrt{\left(\frac{cP}{N}\right)}}{\frac{bP}{N}} = \frac{a}{b}\sqrt{\left(\frac{cP}{N}\right)}$$

$$K = \frac{p_{H_2}\sqrt{(p_{O_2})}}{p_{H_2O}} = \frac{\frac{dP}{N}\sqrt{\left(\frac{cP}{N}\right)}}{\frac{eP}{N}} = \frac{d}{e}\sqrt{\left(\frac{cP}{N}\right)}$$

where P = total pressure of products = 1 atm
N = total number of kmol of products
$= a + b + c + d + e$

These equations can be solved (preferably by computer) and the flame temperature then determined from the energy balance.

1.6 Combustion in a gas turbine

A gas turbine is supplied with fuel of calorific value 20 MJ/kg, and gravimetric analysis 65 % C, 25 % H_2, 10 % O_2. The compressor takes in air at 1 bar, 27 °C and compresses it to a pressure of 4 bar. The exhaust gases heat the air leaving the compressor, before it enters the combustion chamber.

Assuming that the regenerator effectiveness is 78 %, the expansion in the turbine is isentropic, and the analysis of the dry exhaust gas is 6 % CO_2, 1.5 % CO, determine the maximum temperature in the cycle, thermal efficiency of the plant, and specific fuel consumption.

For air $C_p = 1.005$ kJ/kg K, $\gamma = 1.40$.
Combustion products $C_p = 1.15$ kJ/kg K, $\gamma = 1.33$.

Solution Consider 1 kg fuel. The combustion equation can be written, in terms of kmols,

$$\frac{0.65}{12}[C] + \frac{0.25}{2}[H_2] + \frac{0.10}{32}[O_2] + 0.21A[O_2] + 0.79A[N_2]$$

$$\rightarrow a[CO_2] + b[CO] + c[O_2] + d[N_2] + e[H_2O]$$

A carbon balance gives $\frac{0.65}{12} = a + b = 0.0542$

A hydrogen balance gives $0.25/2 = e = 0.125$
An oxygen balance gives $0.10/32 + 0.21A = a + \frac{1}{2}b + c + \frac{1}{2}e$
A nitrogen balance gives $0.79A = d$

Also % $CO_2 = 6 = \frac{100a}{a + b + c + d}$

% CO $= 1.5 = \frac{100a}{a + b + c + d}$

$$\% \, O_2 = \frac{100c}{a + b + c + d}$$

Therefore, $b/a = 1.5/6$, so $0.0542 = a + a/4$, and therefore $a = 0.0434$ and $b = 0.0108$.

Substituting $a + b + c + d = 100a/6 = 0.7233$, but

$$0.10/32 + 0.21A + 0.79A = a + \tfrac{1}{2}b + c + d + \tfrac{1}{2}e$$

therefore, $0.0031 + A = (a + b + c + d) - \tfrac{1}{2}b + \tfrac{1}{2}e = 0.7804$

$$A = 0.7773 \text{ kmol}$$

Hence the air supplied = $0.7773 \times 29 = $ **22.54 kg/kg fuel**.

Referring to Figs 1.1, 1.2, the compressor work (input) = $1.005(T_2 - T_1)$ kJ/kg air, and the turbine work = $1.15(T_3 - T_4)$ kJ/kg gases. The effectiveness of the regenerator,

$$e = \frac{T_5 - T_2}{T_4 - T_2}$$

Therefore, $T_5 = T_2 + e(T_4 - T_2)$.

Now in the compressor $T_2/T_1 = (p_2/p_1)^{\gamma - 1/\gamma} = 4^{0.4/1.4} = 1.486$.

Therefore, $T_2 = 446$ K. In the turbine $T_3/T_4 = (4)^{0.33/1.33} = 1.414$.

An energy balance on the combustion chamber gives

$$m_a C_{pa}(T_2 - T_0) + m_f(CV) = (m_a + m_f)C_{pg}(T_3 - T_0)$$

where T_0 = datum temperature, say 15 °C. Therefore,

$$22.54(1.005)(T_2 - 288) + 1.0(20\,000) = 23.54(1.15)(T_3 - 288)$$

Therefore, $22.54(1.005)(446 - 288) + 20\,000 = 27.071(T_3 - 288)$

$$T_3 = 1159 \text{ K} \quad \text{and} \quad T_4 = 820 \text{ K}$$

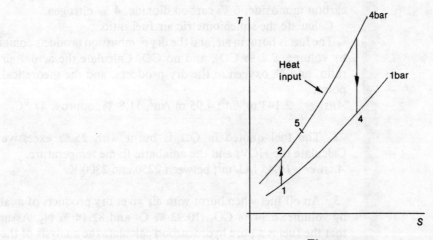

Figure 1.1

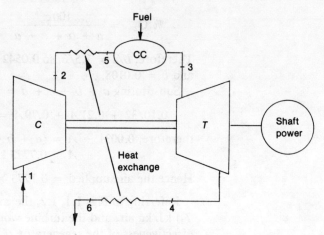

Figure 1.2

The net work output/kg fuel = turbine work − compressor work = −22.54(1.005)(446 − 288) + 23.54(1.15)(1159 − 820) kJ = 9186 − 3579 = 5607 kJ.

Now $T_4 = 1159/1.414 = 820$ K. Therefore,
$$T_5 = 446 + 0.78(820 − 446) = 738 \text{ K}.$$

Therefore, heat input = 23.54(1.15)(1159 − 738) = 11 400 kJ.

Therefore, efficiency of cycle = 5607/11 400 = **0.49**.

Specific fuel consumption = 3600/5607 = **0.64 kg/kWh**.

Problems

1 The analysis of a fuel, by volume, is 50 % hydrogen, 40 % carbon monoxide, 6 % carbon dioxide, 4 % nitrogen.
Calculate the stoichiometric air:fuel ratio.
The fuel is burnt in air, and the dry combustion products contained, by volume, 9.2 % CO_2, and no CO. Calculate the actual air:fuel ratio, and % oxygen in the dry products, and the theoretical dew point.
Answer 2.143 m³/m³; 4.95 m³/m³, 11.8 %; approx. 47 °C.

2 The fuel quoted in Q.1 is burnt with 25 % excessive air. Calculate the HCV, and the adiabatic flame temperature.
Answer 10.45 MJ/m³; between 2250 and 2300 K.

3 An oil fuel when burnt with air gives dry products of analysis, by volume: 6.94 % CO_2, 10.92 % O_2 and 82.14 % N_2. Assuming that the fuel is a pure hydrocarbon calculate the analysis of the fuel (by mass) and % excess air used.

Also calculate the ratio of the actual % CO_2 to the % CO_2 in the stoichiometric dry products.
Answer 83.9 % C, 16.1 % H_2; 100 %; 0.48.

4 The analysis of a fuel, by mass, is 76 % C, 8 % H_2, 8 % H_2O. Calculate the stoichiometric air:fuel ratio, by mass.

The dry combustion products contain, by volume, 10 % CO_2, 7.23 % O_2. Calculate the excess air used, and the % CO in the dry products.

What inferences can be drawn from the presence of the carbon monoxide?
Answer 11.50 kg/kg, 4.27 %, 1.51 %.

5 Ethanol (C_2H_6O) is burnt with (a) 20 % excess air, (b) 20 % air deficiency.

Calculate the analysis of the dry combustion products in each case. Assume in (a) that there is no CO or H_2 in the products, and in (b) no O_2.
Answer (a) 12.39 % CO_2, 3.72 % O_2, 83.89 % N_2;
(b) 7.25 % CO_2, 10.88 % CO, 81.87 % N_2.

6 Heptane (C_7H_{16}) is burnt in an engine, and the rich mixture gave an analysis of the products, by volume, as 8.8 % CO_2, 9.2 % CO, 7.8 % H_2, 0.4 % CH_4 and 73.8 % N_2.

Calculate the air:fuel ratio, and % excess air used.
Briefly explain why CO, H_2, CH_4 appear in the products.
Answer 10.4 kg/kg; 31.6 % def.

7 A gaseous fuel has an analysis, by volume, of 91 % CH_4, 3 % C_2H_6, 2 % C_3H_8, 2 % N_2, 2 % CO_2.

Calculate the stoichiometric air:fuel ratio; volume of wet products/nm^3 fuel; theoretical dew point; and % CO_2 in the dry stoichiometric products. The total pressure of the products = 1 bar.

The fuel is fired in a boiler with 10 % excess air. The flue gases are exhausted at 350 °C against a pressure of 750 N/m² by a fan of efficiency 60 %.

Calculate the volume of wet products and fan input power when the fuel firing rate is 0.1 m³/s.
Answer 9.64 nm³/nm³; 10.68 nm³/nm³; 58 °C; 12.09 %; 1.16 m³/s; 3.32 kW.

8 A stoichiometric mixture of octane and air burns at a pressure of 1 bar to give the combustion equation:

$$C_8H_{18} + 12\tfrac{1}{2}O_2 \rightarrow 7.09CO_2 + 8.80H_2O + 0.91CO + 0.20H_2 + 0.56O_2$$

Calculate the numerical value of the equilibrium constants,

$$K_1 = \frac{p_{CO_2}}{p_{CO}\sqrt{(p_{O_2})}}, \quad K_2 = \frac{p_{H_2O}}{p_{H_2}\sqrt{(p_{O_2})}}$$

Estimate the adiabatic flame temperature, taking the base temperature as 25 °C.

Answer 84, 470; nearly 2500 K.

9 The chemical formula of a fuel is CH_3NO_2. Calculate the stoichiometric air:fuel ratio, kg/kg and the composition of the stoichiometric wet products; and the adiabatic flame temperature.

Heat of reaction = -756 MJ/kg mol.

The dissociated equation is:

$$CH_3NO_2 + \tfrac{3}{4}O_2 \rightarrow aCO_2 + bH_2O + cCO + dH_2 + eO_2 + fN_2 + 2.82N_2$$

At 3600 K, the equilibrium constants are $K_1 = 0.494$, $K_2 = 3.927$. Show that $a = 0.142$ satisfies the requirements of mass balances and equilibrium constants.

Why would the temperature be less than this value in the actual combustion?

Answer 1.7 kg/kg; 3600 K.

10 A fuel oil has an analysis, by mass, of 84.8 % C, 15.2 % H_2.

Calculate the stoichiometric air:fuel ratio (kg/kg).

If the oil is burnt, at 1 bar, with 20 % excess air determine the adiabatic flame temperature. Also determine the adiabatic flame temperature if the oil was burnt with the 20 % excess oxygen.

Answer 15 kg/kg; approx. 2300 K; 6000 K.

11 The analysis of a fuel, by volume, is 97.5 % CH_4, 1.8 % CO_2, 0.7 % N_2.

When burnt with air the dry flue gas analysis, by volume, was 9.7 % CO_2, 3.8 % O_2, 86.5 % N_2.

Calculate the stoichiometric air:fuel ratio, and % excess air used; and the ratio of the stoichiometric % CO_2 to the actual % CO_2. Comment on the value of the ratio.

Answer 9.29 m³/m³; 20.5 %, 1.23.

12 The volumetric analysis of a gaseous fuel is 25 % CH_4, 45 % H_2, 3 % CO_2, 10 % CO, 1 % O_2, 16 % N_2.

Calculate the stoichiometric air:fuel ratio; the analysis of the dry products and volume of wet products/m³ fuel if the fuel is burnt with 25 % excess air.

Answer 3.64 m³/m³; 8.8 % CO_2, 4.4 % O_2, 86.8 N_2; 4.96 m³/m³.

13 Methane, CH_4, is burnt at atmospheric pressure with 95 % stoichiometric air.

Calculate the kmols of CO_2, H_2O, CO, H_2, O_2, N_2 given the following information.

$$K = \frac{p_{CO}p_{H_2O}}{p_{CO_2}p_{H_2}} = 5.36, \qquad \frac{\text{kmol } O_2}{\text{kmol total products}} = 0.002$$

Answer 0.840, 1.931, 0.160, 0.069, 0.014, 7.150.

14 A furnace is fired with a gaseous fuel and 20 % excess air. The volumetric analysis of the fuel is CH_4 20 %, CO 18 %, H_2 50 %, 12 % N_2.

Estimate the calorific value of the fuel and air volume at a firing rate of 10 m³/s.

Answer 42.29 m³/s; 14.8 MJ/nm³.

The fuel is changed to natural gas, which can be taken as methane CH_4. Calculate the % excess air required in this conversion if the heat input in the fuel remains the same, at the same air flow.

Combustion enthalpies: CH_4 = 35.8, CO = 12.6, H_2 = 10.8 MJ/m³.

Answer 7.2 %.

15 A coal analysis by mass is 73 % C, 12 % ash, 15 % H_2O. When burned in a boiler the residue contains 18 % C by mass. The analysis of the dry flue gas, by volume, is CO_2 11.8 %, CO 1.3 %, O_2 5.5 %.

Calculate the % C in the coal which undergoes combustion and air used (kg/kg coal).

Answer 97 %; 13.47.

16 The analysis of a coal by mass is 82 % C, 6 % H_2, 6 % ash, 2 % H_2O. Calculate the stoichiometric air:fuel ratio.

The actual air supplied = 18 kg/kg. Given that 80 % of the carbon is completely burnt and all the hydrogen, calculate the analysis, by volume, of the dry products.

Answer 11.42 kg/kg; 9.1 % CO_2, 2.3 % CO, 7.2 % O_2, 81.4 % N_2.

17 The analysis of a gaseous fuel by volume is 80 % CH_4, 12 % C_2H_6, 8 % N_2.

It is burned with preheated air (enthalpy 0.26 MJ/m³) and the combustion products contain dry, by volume, 7.0 % CO_2 and no CO.

Determine the adiabatic flame temperature, neglecting dissociation.

(Combustion enthalpy: CH_4 = 35.8, C_2H_6 = 63.7 MJ/m³.)

Answer approx. 1850 K.

18 A boiler is fired with dual-fuel burners, using natural gas and fuel oil.

Analysis of natural gas, by volume = 94 % CH_4, 2 % C_2H_6, 4 % CO_2.

Analysis of fuel oil, by mass = 87 % C, 13 % H_2. The analysis of the dry combustion products is 10.00 % CO_2, 0.64 % CO, 4.56 % O_2, 84.80 % N_2.

Calculate the m³ gas used/kg oil.

Answer 2.26.

19 The volumetric analysis of a gaseous fuel, by volume, is 5.6 % CO_2, 7.0 % C_2H_4, 0.4 % O_2, 30.4 % CO, 37.0 % H_2, 14.0 % CH_4, 5.6 % N_2.

Estimate the adiabatic flame temperature if the gas is burned with 3.5 m³ air/m³ fuel, using the data in Table 1.1.

Table 1.1

$T(°C)$	Enthalpy above 0 °C (MJ/m³)			
	N_2	CO	CO_2	H_2O
2100	3.148	3.178	5.129	4.133
2300	3.476	3.506	5.673	4.607
2500	3.808	3.838	6.222	5.084
2700	4.137	4.170	6.763	5.573

Answer approx. 2700 K.

20 A diesel engine is supplied with fuel of calorific value 44 MJ/kg and analysis (by mass) 86 % carbon, 14 % hydrogen. The engine develops 75 kW brake power at a speed of 1800 rev/min, with a brake thermal efficiency of 32 %. Volumetric efficiency at STP is 85 %.

Determine the swept volume required if the fuel is burned with 50 % excess air.

Answer 0.00717 m³.

21 Carbon monoxide is burnt with 50 % excess air. At 2500 K, 3 bar the products contain 7.0 % O_2, 22.55 % CO_2 by volume. Calculate the value of the equilibrium constants for CO_2.

Also estimate the adiabatic flame temperature, neglecting dissociation and neglecting the effect of the pressure.

Answer 27.54 atm$^{-½}$; approx. 2450 K.

22 A hydrocarbon fuel is burned with air and a volumetric analysis of the dry combustion products gave:

CO_2 7 %, CO 5 %, O_2 3 %, N_2 85 %

Determine the gravimetric analysis of the fuel, the air:fuel ratio (kg/kg), and percentage excess air supplied.

Answer 78 % C, 22 % H_2; 16.9 kg/kg; 2.66 %.

23 Explain why it is necessary to operate above the dew point temperature in the exit gas stream from heating plant.

Methane, of lower CV 35.8 MJ/m³, is burned with stoichiometric air at 298 K, and the flue gases leave the system at 600 K. Determine the percentage of the thermal energy in the fuel which is transferred to the flue gas, and the kg moisture condensed/kg of methane burned if the flue gas is cooled to 303 K.

The mean specific heats at constant pressure can be taken as

CO_2 $1.88 \text{ kJ/m}^3 \text{ K}$
H_2O 1.56
N_2 1.32

The flue gas pressure = 1.01 bar.
Answer 12.6 %; 2.25.

24 Explain the significance of the second-law analysis of a chemical reaction. Indicate why this is necessary in order to determine the final products of a combustion process and the energy transfer available from this.

An I.C. engine burns methane, CH_4, as a fuel with 10 % excess air. Construct a set of equations from which it would be possible to calculate the composition of the products of combustion at a point in the expansion process when the pressure is 8 atm, and the temperature is 2330 °C.

Explain why this may not give the same result as an analysis of a sample taken from the engine at this stage.

25 A fuel contains, by volume, 70 % hexane (C_6H_{14}) and 30 % benzene (C_6H_6).

Determine the stoichiometric air:fuel ratio (by mass) for the fuel. Also determine the volumetric analysis of the wet products when the fuel is burned with 20 % excess air, if

(a) all the water vapour is present,
(b) the products are cooled to 1 bar, 15 °C.

Answer 14.7 kg/kg; (a) 11.2 CO_2, 3.3 O_2, 74.7 N_2, 10.8 H_2O;
(b) 12.5 CO_2, 3.7 O_2, 83.8 N_2.

2

Nozzles: rocket propulsion

The flow of a fluid through a nozzle, in which pressure or thermal energy is converted into kinetic energy, is of importance in energy-conversion plant such as steam turbines, gas turbines, hydraulic impulse turbines.

The energy conversion is from chemical energy of the fuel/oxidant into kinetic energy in rocket propulsion.

Convergent–divergent nozzle

The subscripts 0 and e refer to the stagnation (or total) conditions, and exit plane conditions respectively. The superscript asterisk refers to the critical (throat) conditions, and are shown in Fig. 2.1. For isentropic flow of ideal gas the mass flow rate is given by the equation

$$m = A \left(\frac{p}{p_0}\right)^{1/\gamma} \sqrt{\left(\frac{2\gamma p_0 \rho_0}{\gamma - 1}\left[1 - \left(\frac{p}{p_0}\right)^{(\gamma-1)/\gamma}\right]\right)}$$

At the throat $T^*/T_0 = \left(\dfrac{2}{\gamma + 1}\right)$ and is known as the critical temperature ratio. The critical pressure ratio is therefore given by the equation

$$p^*/p_0 = \left(\frac{2}{\gamma + 1}\right)^{\gamma/(\gamma - 1)}$$

In terms of the throat area A^*

$$m = A^* \sqrt{(\gamma p_0 \rho_0)} \times \left(\frac{2}{\gamma + 1}\right)^{(\gamma+1)/2(\gamma-1)}$$

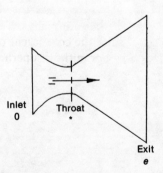

Figure 2.1

The thrust developed is $F = mu_e + (p_e - p_a)A_e$, where p_a is the ambient pressure, and u_e = fluid velocity at the exit plane.

The Mach number M = fluid velocity/local acoustic velocity = $u/\sqrt{\gamma RT}$. At the throat $M^* = 1$.

Steam nozzles

Steam does not behave as an ideal gas, and the enthalpy h and specific volume v are normally used instead of temperature and density.

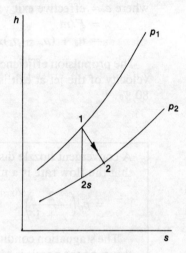

Figure 2.2

Referring to Fig. 2.2, the nozzle efficiency

$$= \Delta h / \Delta h_s = \frac{h_1 - h_2}{h_{1s} - h_{2s}}$$

Application of the steady flow energy equation gives

$$u_2^2 - u_1^2 = 2\Delta h = 2(h_1 - h_2)$$

In many cases the velocity at inlet can be neglected. Also the frictional effect is normally taken to influence the divergent part of the nozzle only, since the convergent part is relatively short.

For initially superheated steam $\gamma = 1.30$, and for initially dry saturated steam $\gamma = 1.135$. These values can be used with the ideal gas laws to give a reasonable approximation.

Rocket propulsion

The performance of a rocket propulsion unit can be described by several parameters. These include:

specific impulse $I_s = F/m \,(\text{Ns/kg})$ or $F/mg \,(\text{s})$
total impulse $I_t = \int F \, dt \,(\text{N})$
specific propellant consumption SPC $= m/F = 1/I_s$

mass ratio = final mass/initial mass m_0
specific power $P_s = \frac{1}{2} m u_e^2 / m_0$ (W/kg)
thrust coefficient $C_f = F/p_0 A^*$

If the velocity of the rocket, in the opposite direction to the gas velocity at exit, is u, the propulsive efficiency is

$$\eta_p = \frac{Fu}{Fu + \frac{1}{2}m(c-u)^2} = \frac{2(u/c)}{1 + (u/c)^2}$$

where c = effective exit velocity with respect to the vehicle
     = F/m
     = $u_e + (p_e - p_a)A_e/m$

The propulsion efficiency is a maximum when $u = u_e$, i.e. the absolute velocity of the jet at exit is zero. In practice the value is between 50 and 80 %.

2.1 Convergent nozzles

A convergent nozzle discharges a gas from a large chamber. Prove that the flow rate is a maximum when the pressure at the throat

$$= p_0 \left(\frac{2}{\gamma + 1} \right)^{\gamma/(\gamma - 1)}$$

The stagnation conditions in the chamber are 20 bar, 2800 K. The thrust on the nozzle is 4500 N when the ambient pressure = 0.18 bar. $\gamma = 1.30$. $R = 360$ J/kg K. Calculate the throat area, velocity of flow and temperature at the throat, and mass flow rate.

Solution Assuming isentropic flow and ideal gas p/ρ^γ = constant, $p = \rho RT$, and the steady-flow energy equation is $h_0 = h + \frac{1}{2}u^2$ = constant, where the enthalpy $h = C_p T$. Also $C_p = R/(\gamma - 1)$.

$$C_p T_0 = C_p T + \frac{1}{2} u^2$$

Therefore $u^2 = 2C_p(T_0 - T) = \dfrac{2\gamma R}{\gamma - 1}(T_0 - T)$

$$= \frac{2\gamma}{\gamma - 1} \frac{p_0}{\rho_0} \left[1 - \frac{p}{\rho} \frac{\rho_0}{p_0} \right]$$

$$= \frac{2\gamma}{\gamma - 1} \frac{p_0}{\rho_0} \left[1 - \left(\frac{p}{p_0} \right)^{1 - 1/\gamma} \right]$$

Therefore $m = \rho A u = \rho_0 A u \left(\dfrac{\rho}{\rho_0} \right) = \rho_0 A u \left(\dfrac{p}{p_0} \right)^{1/\gamma}$

$$= A \left(\frac{p}{p_0} \right)^{1/\gamma} \sqrt{\left(\frac{2\gamma}{\gamma - 1} (p_0 \rho_0) \left[1 - \left(\frac{p}{p_0} \right)^{(\gamma-1)/\gamma} \right] \right)}$$

The mass flow rate is a maximum, for a given area, when $\dfrac{dm}{dr} = 0$, where $r = p/p_0$. Hence

$$\frac{d}{dr}\left[r^{2/\gamma}(1 - r^{(\gamma-1)/\gamma})\right] = 0$$

Therefore $\dfrac{d}{dr}\left[r^{2/\gamma} - r^{(\gamma+1)/\gamma}\right] = 0$

$$\frac{2}{\gamma}r^{2/\gamma - 1} - \left(\frac{\gamma + 1}{\gamma}\right)r^{[(\gamma+1)/\gamma] - 1} = 0$$

$$r^{2/\gamma - 1} = \left(\frac{\gamma + 1}{2}\right)r^{1/\gamma}$$

$$r^{1/\gamma - 1} = \frac{\gamma + 1}{2}$$

$$r = \left(\frac{\gamma + 1}{2}\right)^{1 - 1/\gamma} = \left(\frac{2}{\gamma + 1}\right)^{\gamma/(\gamma - 1)}$$

In this problem $p_0 = 20$ bar, $\gamma = 1.30$, therefore

$$p^* = 20\left(\frac{2}{2.3}\right)^{1.3/0.3} = 10.915 \text{ bar}$$

Also $\dfrac{T^*}{T_0} = \left(\dfrac{p^*}{p_0}\right)^{\gamma - 1/\gamma} = \dfrac{2}{\gamma + 1}$, therefore

$$T^* = \frac{2}{2.3}(2800) = \mathbf{2435 \text{ K}}$$

$$u^* = \sqrt{\left[\frac{2\gamma R}{\gamma - 1}(T_0 - T^*)\right]} = \sqrt{\left[\frac{2(1.3)(360)}{0.3}(2800 - 2435)\right]}$$

$$= \mathbf{1067 \text{ m/s}}$$

The mass flow rate is determined from the thrust

$$F = mu^* + (p^* - p_a)A^*$$

Therefore, $4500 = m(1067) + 10^5(10.915 - 0.18)A^*(\text{N})$
$\qquad\qquad = 1067m + 10.735 \times 10^5 A^*$

Also $m = \rho^* A^* u^* = \dfrac{p^*}{RT^*}A^* u^* = \dfrac{10.915 \times 10 \times 1067 A^*}{360 \times 2435}$

$$= 1328.6 A^*$$

Therefore $4500 = 1067m + \dfrac{10.735 \times 10^5 m}{1328.6}$, giving

$$m = \mathbf{2.40 \text{ kg/s}}$$
$$A^* = \mathbf{0.001\,81 \text{ m}^2}$$

2.2 Rocket performance

> The stagnation conditions in the combustion chamber of a rocket motor are 25 bar, 3520 K. The nozzle area = 0.1 m². The exit area is such that the exit and ambient pressures are equal to 0.012 bar.
> $\gamma = 1.22$, $R = 520$ J/kg K.
>
> Calculate the exit area, gas velocity at the exit, the thrust developed, specific impulse and thrust coefficient, and Mach number at the exit.
>
> Briefly outline the effect of the ambient pressure (a) decreasing, (b) increasing.

Solution At the nozzle throat $p^*/p_0 = \left(\dfrac{2}{\gamma + 1}\right)^{\gamma/(\gamma-1)}$; therefore

$$p^* = 25\left(\frac{2}{2.22}\right)^{5.545} = 14.016 \text{ bar}$$

$$T^* = T_0\left(\frac{2}{\gamma + 1}\right) = 3520\left(\frac{2}{2.22}\right) = 3171 \text{ K}$$

Hence, $\rho^* = p^*/RT^* = 0.850$ kg/m³, $u^* = \sqrt{(\gamma RT^*)} = 1418$ m/s; therefore $m = \rho^* A^* u^* = 120.6$ kg/s.

For the exit plane $u_e^2 = \dfrac{2\gamma R}{\gamma - 1}(T_0 - T_e)$

where $T_e = T_0\left(\dfrac{p_e}{p_0}\right)^{(\gamma-1)/\gamma} = 3520\left(\dfrac{0.012}{25}\right) = 887$ K

Therefore $u_e = \sqrt{\left[\dfrac{2 \times 1.22 \times 520}{0.22}(3520 - 887)\right]} = \mathbf{3897 \text{ m/s}}$

$\rho_e = p_e/RT_e = 0.012 \times 10^5/520 \times 887 = \mathbf{0.0026 \text{ kg/m}^3}$

Therefore $A_e = m/\rho_e u_e = 120.6/(0.0026 \times 3897) = \mathbf{11.903 \text{ m}^2}$

Thrust $F = mu_e$ (since $p_e = p_a$) = **469 978 N**

Specific impulse $I_s = F/m = $ **3897 N s/kg**

Thrust coefficient $C_f = F/p_0 A^* = \dfrac{469\,978}{25 \times 10 \times 0.1} = \mathbf{1.88}$

Acoustic velocity $a_e = \sqrt{(\gamma RT_e)} = \sqrt{(1.22 \times 520 \times 887)} = \mathbf{750 \text{ m/s}}$

Therefore, $M_e = u_e/a_e = 3897/750 = \mathbf{5.20}$

(a) If the ambient pressure decreases so that $p_e < p_a$, the nozzle is said to be over-expanded. This would cause the supersonic gas flow to expand outside the nozzle, and a pattern of oblique shock waves would be set up in the jet leaving the nozzle.

(b) If the ambient pressure increases, so that $p_e > p_a$, the nozzle is under-expanded. In this case a normal shock wave would be set up in the divergent portion of the nozzle, decelerating the flow.

2.3 Convergent–divergent nozzle

> The stagnation conditions in a rocket combustion chamber are 3000 K, 30 bar. The pressure at the exit plane = 0.5 bar, and the ambient pressure = 0.05 bar. The mass flow rate of gas = 12 kg/s. The forward velocity of the vehicle is 1500 m/s.
> $\gamma = 1.25$. $R = 500$ J/kg K.
> Assuming that the flow is isentropic, calculate the throat and exit areas, specific impulse, thrust coefficient and propulsive efficiency.

Solution The conditions at the throat are determined

$$p^* = p_0 \left(\frac{2}{\gamma + 1}\right)^{\gamma/(\gamma - 1)} = 30\left(\frac{2}{2.25}\right) = 16.65 \text{ bar}$$

$$T^* = T_0(2/(\gamma + 1)) = 3000(2/2.25) = 2667 \text{ K}$$

$$\rho^* = \frac{16.65 \times 10^5}{500 \times 2667} = 1.2486 \text{ kg/m}^3$$

$$u^* = \sqrt{\left[\frac{2\gamma R}{\gamma - 1}(T_0 - T^*)\right]} = \sqrt{\left[\frac{2(1.25)(500)}{0.25}(3000 - 2667)\right]}$$

$$= 1290 \text{ m/s}$$

$$A^* = m/\rho^* u^* = 12/(1290 \times 1.2486) = \mathbf{0.00745 \ m^2}$$

At the exit plane $\dfrac{p_e}{p_0} = \left(\dfrac{T_e}{T_0}\right)^{\gamma/(\gamma - 1)}$; therefore

$$T_e = T_0 \left(\frac{p_e}{p_0}\right)^{(\gamma - 1)/\gamma}$$

Therefore $T_e = 3000(0.5/30)^{0.2} = 1323$ K

$$\rho_e = p_e/RT_e = \frac{0.5 \times 10^5}{500 \times 1323} = 0.0756 \text{ kg/m}^3$$

Therefore $A_e = 12/(0.0756 \times 2896) = \mathbf{0.0548 \ m^2}$

since $u_e = \sqrt{\left[\dfrac{2(1.25)(500)}{0.25}(3000 - 1323)\right]} = 2896$ m/s

Thus, the throat area = 7.45 cm², exit area = 548 cm².

Thrust $F = mu_e + (p_e - p_a)A_e$
$= 12(2896) + 10^5(0.5 - 0.05)(0.0548)$ N
$= 37218$ N

Specific impulse = F/m = **3102 Ns/kg** or **316 s**

Thrust coefficient = $F/p_0 A^*$ = **1.665**

Propulsive efficiency = $\dfrac{2(u/c)}{1 + (u/c)}$ where $c = F/m = 3102$ m/s

Therefore, $\eta_p = \dfrac{2(0.4836)}{1 + (04836)} = \mathbf{0.784}$

2.4 Steam nozzle

Steam at 30 bar, 350 °C expands through a convergent–divergent nozzle. The pressure at the exit plane = 3 bar. Flow rate = 0.5 kg/s. Nozzle efficiency = 80 %. Assuming that the velocity at inlet is negligible, determine the throat and exit areas, steam velocity at the exit, and the steam condition at the exit plane. The critical pressure ratio can be taken as 0.546, corresponding to $\gamma = 1.30$.

Solution At the inlet pressure of 30 bar, the saturation temperature is 234 °C, therefore the steam is superheated, and $h_1 = 3117$ kJ/kg, $s_1 = 6.744$ kJ/kg K.

At the throat $p^* = 0.564 p_0 = 16.4$ bar. Assuming that the flow in the convergent portion is isentropic $s^* = s_1 = 6.744$ kJ/kg K, therefore the steam is superheated, and $h^* = 2965$ kJ/kg.

$$u^* = \sqrt{[2(h_1 - h^*)]} = \sqrt{[2 \times 10^3 (3117 - 2965)]} = 551.4 \text{ m/s}$$

The specific volume $v^* = 0.1407$ m^3/kg. Therefore,

$$A^* = mv^*/u^* = 0.5(0.1407)/551.4 = 1.276 \times 10^{-4} \text{ m}^2$$

Therefore throat area = **127.6 mm^2**.

In the divergent portion there are frictional effects. Considering isentropic flow

$$s_{2s} = s^* = 6.744 \text{ kJ/kg K} \quad \text{and} \quad p_2 = 3 \text{ bar}$$

At this pressure $s_g = 6.993$ kJ/kg K; therefore the steam is wet.

$$6.744 = s_f + x_{2s} s_{fg} = 1.672 + 5.321 x_{2s}$$

therefore $x_{2s} = 0.953$

$$h_{2s} = h_f + x_{2s} h_{fg} = 561 + 0.953(2164) = 2624 \text{ kJ/kg}$$

therefore isentropic enthalpy drop $\Delta h_s = 2965 - 2624 = 341$ kJ/kg, therefore, actual enthalpy drop $\Delta h = 0.8 \Delta h_s = 273$ kJ/kg. Hence $h_2 = 2965 - 273 = 2692$ kJ/kg.

$$x_2 = \frac{2692 - 561}{2164} = \mathbf{0.985}$$

Also, $v_2 = 0.985$, $v_g = 0.985(0.6057) = 0.5965$ m^3/kg, and

$$u_2^2 - u^{*2} = 2\Delta h = 2 \times 10^3 \times 273$$

therefore, $u_2 = \mathbf{922.0 \text{ m/s}}$

giving $A_2 = 0.5(0.5965)/922 = 3.235 \times 10^{-4}$ m^2; therefore exit area = **323.5 mm^2**.

2.5 Steam nozzle

Dry saturated steam at 10 bar is expanded in a convergent–divergent nozzle. The velocity of the steam at exit = 685 m/s. Flow rate = 7 kg/s. Nozzle efficiency = 85 %. Critical pressure ratio = 0.58.
Determine the throat and exit areas, and the pressure at the nozzle exit.

Solution Assuming that the flow is isentropic up to the throat $p^* = 0.58 p_1 = 5.80$ bar, and $h_1 = 2778$ kJ/kg, $s_1 = s^* = 6.586$ kJ/kg K. Therefore $6.586 = 1.917 + 4.855 x^*$; therefore $x^* = 0.962$.

$$h^* = 664 + 0.962(2091) = 2676 \text{ kJ/kg}$$
$$u^* = \sqrt{[2(h_1 - h^*)]} = 451.7 \text{ m/s}$$
$$v^* = 0.962(0.3264) = 0.3140 \text{ m}^3/\text{kg}.$$

Therefore $A^* = 7(0.3140)/451.7 = 0.00487$ m^2; therefore

throat area = **4870 mm^2**.

From the throat to the exit

$$u_2^2 = u^{*2} = 2(h^* - h_2)$$

Therefore $(685) - (451.7)^2 = 2 \times 10^3 (2676 - h_2)$

$h_2 = 2543$ kJ/kg.

The actual enthalpy drop $\Delta h = 2676 - 2543 = 133$ kJ/kg; therefore the isentropic drop $\Delta h_s = 133/0.85 = 156$ kJ/kg. Therefore if the expansion was isentropic, $h_{2s} = 2676 - 156 = 2520$ kJ/kg. Also the entropy would be $s_{2s} = s^* = 6.586$ kJ/kg K. The steam pressure is

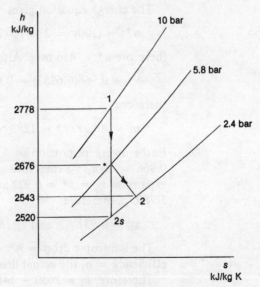

Figure 2.3

determined by the requirement that the enthalpy is 2520 kJ/kg and the entropy 6.586 kJ/kg K. Using the steam tables gives $p_2 = $ **2.4 bar**.

The actual steam condition is $h_2 = 2543$ kJ/kg at 2.4 bar pressure; therefore $2543 = 530 + 2185x_2$, therefore $x_2 = 0.921$ and $v_2 = 0.921(0.7466) = 0.6876$ kg/m³, therefore

$$A_2 = 7(0.6876/685) = 0.0070 \text{ m}^2$$

therefore, exit area = **7000 mm²**.

This problem can be solved more easily, but less accurately, by the use of a Molier (h–s) diagram. Referring to Fig. 2.3 the initial point 1 is located at 10 bar, dry saturated. A constant-entropy line down to a pressure of 5.8 bar locates the throat conditions, and continuing this line to a point where $h = 2520$ kJ/kg locates point 2s, and hence the pressure p_2.

2.6 Steam nozzle

> Dry saturated steam at 5 bar enters a convergent-divergent nozzle at a velocity of 100 m/s. The exit pressure = 1.5 bar, and the throat and exit areas are 1280 and 1600 mm² respectively. Critical pressure ratio = 0.58.
> Estimate the mass flow rate and nozzle efficiency.

Solution At the nozzle inlet $h_1 = 2749$ kJ/kg, $s_1 = 6.822$ kJ/kg. At the throat $p^* = 0.58$, $p_1 = 2.9$ bar. Since

$$s^* = s_1, \quad 6.822 = 1.660 + 5.344x^*$$

$x^* = 0.966$; therefore

$$h^* = 556 + 0.966(2168) = 2650 \text{ kJ/kg}.$$

The energy equation gives

$$u^{*2} - (100)^2 = 2 \times 10^3(2749 - 2650)$$

therefore $u^* = 456$ m/s. Also

$$v^* = 0.966(0.6253) = 0.6040 \text{ m}^3/\text{kg}$$

therefore

$$m = A^*u^*/v^* = 1280 \times 10^{-6} \times 456/0.6040 = \textbf{0.966 kg/s}.$$

In the divergent portion $u_2^2 - (456)^2 = 2 \times 10^3(2650 - h_2)$. Also $m = 0.966 = A_2 u_2/v_2$; therefore $u_2/v_2 = 0.966/A_2 = 603.8$. For isentropic expansion $s_{2s} = s^* = 6.822$ at $p_2 = 1.5$ bar; therefore $6.822 = 1.434 + 5.789 x_{2s}$, therefore

$$x_{2s} = 0.931 \quad \text{and} \quad h_{2s} = 467 + 0.931(2226) = 2539 \text{ kJ/kg}.$$

The isentropic drop = $h^* - h_{2s} = 148$ kJ/kg; therefore if the nozzle efficiency = η, the actual drop is 148η.

Therefore, $h_2 = 2650 - 148\eta$ kJ/kg.

The equations to be solved are

$$h_2 = 2650 - 148\eta$$
$$u_2^2 = 207\,936 + 2000(2650 - h_2)$$
$$u_2 = 603.8 v_2$$

at a pressure $p_2 = 1.5$ bar.

These have to be solved by iteration since the specific volume *cannot* be determined from the ideal gas laws.

A value for the dryness fraction x_2 is assumed, v_2 determined from tables, u_2 calculated. This is compared in Table 2.1 with the value of u_2 determined from the enthalpy h_2. At 1.5 bar, $v_g = 1.159$ m³/kg, $h_f = 467$, $h_{fg} = 2226$ kJ/kg.

Table 2.1

x_2	v_2	h_2	$u_2 = 603.8 v_2$	$u_2 = \sqrt{[207\,936 + 2000(2650 - h_2)]}$
0.94	1.0895	2559	657.8	624.4
0.95	1.1011	2582	664.8	587.0
0.93	1.0779	2537	650.8	658.7
0.938	1.0871	2555	656.4	630.8
0.931	1.0790	2539	651.5	655.7
0.932	1.080	2542	652.2	651.7

From Table 2.1 $x_2 = 0.932$, $h_2 = 2542$

Hence $h_2 = 2650 - 148\eta$; therefore $\eta = \dfrac{2650 - 2542}{148} = \mathbf{0.73}$.

2.7 Rocket motor

Show that for isentropic flow in a convergent–divergent nozzle, expanding fully from an inlet stagnation pressure p_0 to an exit pressure p_e, the ratio of the exit area to throat area is given by

$$\frac{A_e}{A^*} = \frac{k^a \, r^{-1/\gamma}}{\left[\dfrac{\gamma + 1}{\gamma - 1} (1 - r^b) \right]^{1/2}}$$

where $k = 2/(\gamma + 1)$
$r = p_e/p_0$
$a = 1/(\gamma - 1)$
$b = (\gamma - 1)/\gamma$

A rocket motor nozzle expands gases from stagnation conditions, in the combustion chamber, of 35 bar, 3330 K. The exit pressure = 1.0 bar; ambient pressure = 0.1 bar; molecular wt. of gas = 26; $\gamma = 1.25$.

If the throat area = 0.065 m² calculate the exit area, mass flow rate, and thrust, and also the specific impulse and thrust coefficient.

Solution $u^{*2} = 2C_p(T_0 - T^*) = 2C_p T_0(1 - T^*/T_0)$

$$= 2C_p T_0\left(1 - \frac{2}{\gamma + 1}\right)$$

$$= 2C_p T_0\left(\frac{\gamma - 1}{\gamma + 1}\right)$$

$$p^* = p_0\left(\frac{2}{\gamma + 1}\right)^{\gamma/(\gamma-1)} = p_0 k^{1/b}, \rho^* = p^*/RT^* = \rho_0 k^a$$

Therefore, $A^* = \dfrac{m}{\rho^* u^*} = \dfrac{m}{\rho_0 k^a \sqrt{2\left(\dfrac{\gamma - 1}{\gamma + 1}\right) C_p T_0}}$

$$u_e = \sqrt{2C_p(T_0 - T_e)} = \sqrt{2C_p T_0\left[1 - \left(\frac{p_e}{p_0}\right)^{(\gamma-1)/\gamma}\right]} = \sqrt{2C_p T_0(1 - r^b)}$$

$\rho_e = \dfrac{p_e}{RT_e} = \dfrac{p_e}{RT_0 r^b}$, therefore $A_e = \dfrac{m}{\dfrac{p_e}{RT_0 r^b}\sqrt{2C_p T_0(1 - r^b)}}$

Substituting $\dfrac{A^*}{A_e} = \dfrac{\dfrac{p_e}{RT_0 r^b} \cdot 2C_p T_0(1 - r^b)}{\rho_0 k^a 2\left(\dfrac{\gamma - 1}{\gamma + 1}\right) C_p T_0}$

$$= \dfrac{r\dfrac{p_0}{RT_0} r^{-b}\sqrt{(1 - r^b)}}{\rho_0 k^a\left(\dfrac{\gamma - 1}{\gamma + 1}\right)}$$

$$\dfrac{A_e}{A^*} = \dfrac{r^b k^a\sqrt{\left(\dfrac{\gamma - 1}{\gamma + 1}\right)}}{r\sqrt{(1 - r^b)}} = \dfrac{k^a r^{-1/\gamma}}{\sqrt{\left[\left(\dfrac{\gamma + 1}{\gamma - 1}\right)(1 - r^b)\right]}}$$

since $r^b/r = r^{b-1} = r^{-1/\gamma}$.

In the problem given $\gamma = 1.25$, therefore $k = 0.8889$, $a = 4$, $b = 0.20$, and $r = p_e/p_0 = 1/35$.

Therefore, $\dfrac{A_e}{A^*} = \dfrac{(0.8889)\left(\dfrac{1}{35}\right)^{-0.8}}{\left(\dfrac{2.25}{0.25}\left[1 - \left(\dfrac{1}{35}\right)^{0.2}\right]\right)^{1/2}} = \dfrac{10.7316}{2.1400} = 5.0146$

Therefore, $A_e = 5.0146(0.065) = \mathbf{0.326 \ m^2}$.

At the throat $M^* = 1$; therefore

$$u^* = \sqrt{(\gamma RT^*)} = \sqrt{\left[1.25\left(\frac{8314}{26}\right) \times \frac{2}{2.25} \times 3330\right]}$$

$$= 1088 \text{ m/s}$$

Also, $p^* = p_0 \left(\dfrac{2}{\gamma + 1}\right)^{\gamma/(\gamma-1)} = 35 \left(\dfrac{2}{2.25}\right) = 19.42 \text{ bar}$

$$\rho^* = \dfrac{19.42 \times 10^5}{\dfrac{8314}{26} \times \dfrac{2}{2.25} \times 3330}$$

$$= 2.052 \text{ kg/m}^3$$

Therefore $m = \rho^* A^* u^* =$ **145.1 kg/s**.

Thrust $F = mu_e + (p_e - 5p_a)A_e$
$= 145.1 u_e + 10(1.0 - 0.1)(0.326) \text{ N}$

The exit velocity u_e can be determined from the energy equation

$$u_e = \sqrt{[2C_p(T_0 - T_e)]}$$

where $C_p = \dfrac{R}{\gamma - 1} = \dfrac{1.25}{0.25} \times \dfrac{8314}{26}$

$= 1599 \text{ J/kg K}$, and $T_e = T_0 \left(1 - \left(\dfrac{p_e}{p_0}\right)^{(\gamma-1)/\gamma}\right)$

$= 3330(1 - 0.4911) = 1695 \text{ K}$

Hence $u_e = 2 \times 1599 \times (3330 - 1695) = 2287$ m/s. Substituting, we obtain $F =$ **361 184 N**.

The specific impulse $= F/m =$ **2489 Ns/kg**.
The thrust coefficient $= F/(p_0 A^*) =$ **1.588**.

Problems

1 Show that the maximum mass flow rate through a convergent nozzle passing *air* is given by

$$m = 4.04 \times 10^{-3} \dfrac{Ap_0}{\sqrt{T_0}} \text{ kg/s}$$

where A = nozzle area (mm^2), p_0 = reservoir pressure (bar), T_0 = reservoir temperature (K)

Air is discharged from a large container through a convergent nozzle of 10 mm diameter. The conditions in the container are 10 bar, 20 °C. The pressure at the nozzle = 6 bar. Calculate the mass flow rate, and flow Mach number at the nozzle.
Answer 0.184 kg/s; 0.89

2 A container of volume 0.8 m^3 contains air at 150 °C, 200 bar.

The air is discharged through a nozzle of area 150 mm² into ambient air at 1 bar.

Assuming isentropic flow, calculate the time taken for the pressure in the container to fall to 5 bar.
Answer   20.8 s

3   Air flows through a convergent–divergent nozzle from a reservoir in which the conditions are 4 bar, 20 °C. The mass flow rate = 1.0 kg/s. Exit area = 5500 mm². Assuming isentropic flow determine the pressure at the exit, and air velocity at the exit.

Also calculate the thrust on the nozzle if the ambient air pressure = 1 bar.
Answer   1.32 bar, 84.2 m/s; 260 N

4   A convergent–divergent nozzle discharges a gas from a large container, in which the stagnation conditions are 5.5 bar, 760 °C. The pressure at the exit is 1.45 bar, and the temperature 520 °C. $R = 290$ J/kg K, $\gamma = 1.33$.

Calculate the ratio of the exit area to the throat area, and nozzle efficiency.

Assume that the flow is isentropic in the convergent portion of the nozzle.
Answer   1.44; 0.82

5   Air flows through a convergent–divergent nozzle at a rate of 1.2 kg/s. The flow is isentropic, and the stagnation temperature = 20 °C.

At the nozzle exit the pressure = 0.14 bar and the Mach number = 2.8. $C_p = 1.005$ kJ/kg K.

Calculate the throat and exit areas, the stagnation and critical pressures, and temperature and velocity at the exit.
Answer   0.00134, 0.00468 m²; 3.79, 2.00 bar; 114 K, 599 m/s

6   A convergent nozzle is attached to the end of a circular pipe, 10 cm diameter, by a bolted flange. Nozzle diameter = 5 cm. Water is discharged from the nozzle at a velocity of 30 m/s.

Calculate the force on the flange bolts caused by the water flow, and indicate the direction of the force. The water pressure at the flange is 4.5 bar (absolute).
Answer   1424 N; compression

7   A convergent–divergent nozzle passes a gas, for which $\gamma = 1.33$ and $R = 209$ J/kg K. The stagnation conditions at entry are 5.5 bar, 760 °C.

Given that the conditions at the nozzle exit are 524 °C, 1.45 bar, determine the ratio of the exit area to throat area, and nozzle efficiency. The flow may be considered as isentropic in the convergent portion.
Answer   1.46; 61 %

8 In a rocket combustion chamber the stagnation conditions are 25 bar, 3520 K. The throat area = 0.1 m². The exit pressure and ambient pressure are 0.012 bar. γ = 1.22. The molecular weight of the gases = 16.

Calculate the exit area, the Mach number at the exit, the thrust, and specific impulse.

Answer 11.90 m²; 5.18; 469 kN; 3895 N s/kg

9 A rocket operates at an altitude of 12.2 km, where the ambient pressure = 0.18 bar. The chamber stagnation conditions are 2780 K, 20.5 bar.

Given that the thrust developed = 45 000 N, calculate the throat and exit areas, velocity at the throat and exit planes, the temperature and Mach number at the exit, and specific impulse.

γ = 1.30. R = 280 J/kg K.

Answer 0.0138, 0.1455 m²; 938, 2118 m/s; 932 K, 3.64; 2118 N s/kg

10 The stagnation conditions in a rocket combustion chamber are 20 bar, 2900 K. Ambient pressure = 1 bar. The thrust developed is 1350 N, and the specific impulse = 2242 N s/kg.

γ = 1.23. Molecular weight of gases = 22.

Calculate the nozzle throat and exit areas, the velocity at the exit; and show that the nozzle is fully expanded.

Answer 4.78×10^{-4}, 1.68×10^{-3} m²; 2242 m/s

11 A solid propellant rocket develops a thrust of 5 kN against an ambient pressure of 1 bar, and the effective exhaust velocity = 2500 m/s. The exit plane area = 0.01 m².

Determine the thrust developed in a vacuum, and the new effective exhaust velocity.

The burning time of the rocket was 5 s. Calculate the initial mass of propellant.

Answer 6000 N, 3000 m/s, 10 kg

12 A rocket operating at a chamber stagnation pressure of 70 bar produces a mean thrust of 20 kN for a period of 10 s, and in that time consumes 100 kg of propellant. The exit velocity = 1970 m/s.

Calculate the exit area, specific impulse and thrust coefficient, pressure at the exit plane, and nozzle efficiency.

The combustion chamber stagnation temperature = 3000 K.

γ = 1.30. R = 280 J/kg K.

Ambient pressure = 0.1 bar.

Answer 0.017 m², 2000 N s/kg, 1.43; approx. 1.18 bar; 0.87

13 A convergent–divergent nozzle discharges 0.07 kg/s of steam into a chamber, in which the pressure is 1.4 bar. The steam is supplied to the nozzle at 7.33 bar, 200 °C.

Calculate the throat and exit areas of the nozzle. Nozzle efficiency = 90 %. The inlet velocity is negligible.
Answer 6.59×10^{-5}, 10.86×10^{-5} m^2

14 Steam passes through a convergent–divergent nozzle from a pressure of 8 bar. The steam is initially dry saturated ($n = 1.135$). The nozzle efficiency is 90 %. Given that the exit area = 2 × throat area, determine the pressure at the exit.
The inlet velocity is negligible.
Answer approx. 1.2 bar

15 A convergent–divergent nozzle receives dry, saturated steam and discharges it at a velocity of 800 m/s into a chamber at a pressure of 1.4 bar. Nozzle efficiency = 85 %.
Estimate the pressure of the steam supply.
For steam initially dry $n = 1.135$.
Given that the mass flow rate = 10 kg/s, determine the throat and exit areas.
Answer approx. 13 bar, 0.0054, 0.133 m^2

16 Steam at 10 bar, 95 % dry expands isentropically through a nozzle to a pressure of 0.8 bar.
Assuming that the expansion follows the law pv^n = constant, determine the value of n.
Hence calculate the pressure and steam velocity at the throat, and determine the steam temperature at the throat.
Given that the throat area = 30 mm^2 calculate the mass flow rate and the exit area.
Answer 1.13; 5.80 bar, 443 m/s, 158 °C, 0.044 kg/s; 84.5 mm^2

17 Steam, initially dry saturated at 26 bar, expands isentropically in a nozzle to 12 bar. Determine the mass flow rate per cm^2 of throat area and the condition of the steam at the nozzle exit if the expansion is assumed to be

(a) in equilibrium, $n = 1.135$
(b) supersaturated, $n = 1.3$

Briefly explain how nozzle losses can be taken into account in the calculations.
Answer (a) 0.37 kg/s; 93.5 % dry, (b) 0.368 kg/s; superheated

3

Vapour cycles

A vapour cycle is of importance in a large field of energy conversion involving steam plant, in which the chemical energy of the fuel is converted into thermal energy in a boiler. The thermal energy is then converted into mechanical/electrical energy by expanding through a turbine. The cycle is also of importance in the areas of refrigeration and heat pumps, where energy is used to *transfer* heat from one space to another.

A vapour does not behave as an ideal gas so that the vapour properties (pressure, saturation temperature, specific volume, enthalpy, and entropy) must be obtained from vapour property tables or charts.

Carnot and Rankine cycle

Referring to Fig. 3.1 the Carnot cycle consists of two isothermal and two isentropic processes.

The cycle efficiency is

$$\eta_c = 1 - T_2/T_1$$

Net work $= [(h_1 - h_2) - (h_4 - h_3)]/\text{kg}$.

Specific steam consumption SSC = kg/s steam required to generate 1 kWh of work output $= \dfrac{3600}{\text{net work (kJ/kg)}}$.

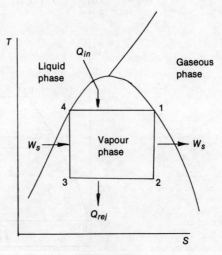

Figure 3.1

VAPOUR CYCLES 33

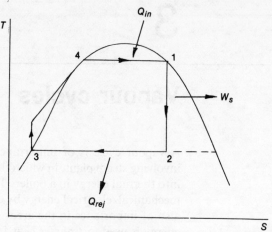

Figure 3.2

The Rankine cycle is shown in Fig. 3.2. In this cycle the vapour is condensed to the saturated liquid state, and the feed pump work is normally negligible. The cycle efficiency is

$$\eta_R = \frac{h_1 - h_2}{h_1 - h_3}$$

neglecting feed pump work (Section 3.1).

The practical cycle requires superheat to avoid wet steam at the turbine exhaust, so avoiding damage to the L.P. blades.

Reheat and regenerative feed heat cycle

The reheat cycle is shown in Fig. 3.3. The steam is expanded in the first stages from 1 to 2, reheated to 3, and then expanded in the second stages from 3 to 4. In this case the net work $= (h_1 - h_2) + (h_3 - h_4)$, and the heat input $= (h_1 - h_5) + (h_3 - h_2)$.

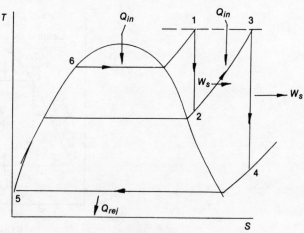

Figure 3.3

34 SOLVING PROBLEMS IN APPLIED THERMODYNAMICS AND ENERGY CONVERSION

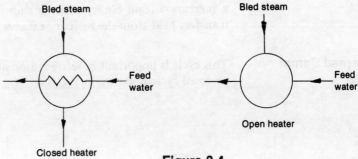

Figure 3.4

The heat supplied from 5 to 6 is below the maximum cycle temperature. If the feed water temperature were increased reversibly by the use of an ideal heat exchanger the cycle efficiency would become equal to the Carnot cycle efficiency. This regenerative heating cycle is approximated by the use of feed heaters, as shown in Fig. 3.4, where a fraction of the total steam flow is bled off at intervals and passed to a feed heater.

Binary vapour cycle This cycle used two vapours, each being the working fluid in a Rankine cycle, and the cycles connected by a heat exchanger. Referring to Fig. 3.5,

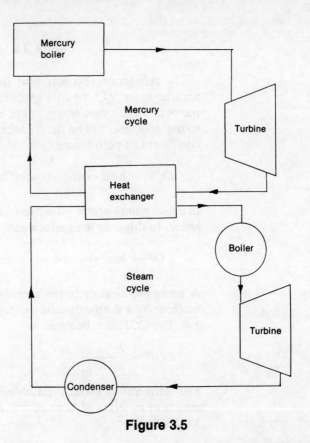

Figure 3.5

VAPOUR CYCLES 35

a mercury–steam binary plant is shown, in which the heat exchanger transfers heat from the mercury exhaust to the feed water.

Reversed Carnot cycle

This cycle is important in refrigeration and heat-pump calculations. The reversed cycle is shown in Fig. 3.6.

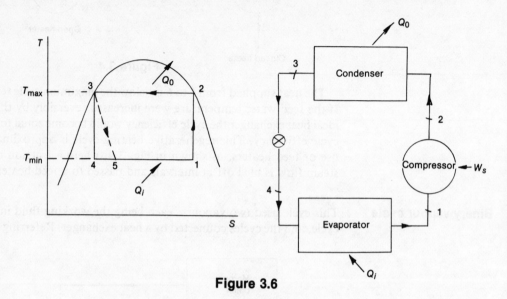

Figure 3.6

The refrigerator extracts heat from a space, Q_i, and rejects heat to another space, Q_o: the cold space temperature is maintained by the heat removal. In this case energy is not being converted from heat into work, so that efficiency has no significance. The performance is measured by the coefficient of performance

$$COP = \text{heat extracted/work input} = \frac{T_{min}}{T_{max} - T_{min}}$$

In a heat pump, heat is extracted from one space to supply heat to another space. In this case the performance is measured by

$$COP = \text{heat rejected/work input} = \frac{T_{max}}{T_{max} - T_{min}}$$

A more practical cycle replaces the expansion cylinder of the Carnot machine by a simple throttle valve, shown by the dotted line 3–5 in Fig. 3.6. The *COP* now becomes

$$COP = \frac{h_1 - h_5}{h_2 - h_1}$$

The refrigerating effect (heat absorbed by the evaporator) is sometimes measured in terms of 'tons of refrigeration'. 1 tonne of refrigeration = 3517 W.

3.1 Carnot and Rankine cycle (steam)

Steam at 100 bar, dry saturated, enters a turbine and expands to a pressure of 0.07 bar. Calculate the Carnot and Rankine cycle efficiencies, and specific steam consumption. Show that in the Rankine cycle the feed-pump work is negligible.

Given that the steam at inlet is superheated to 400 °C, determine the efficiencies and SSC when the turbine isentropic efficiency is (a) 100 %, (b) 80 %.

Solution Referring to Fig. 3.2, $h_1 = 2725$ kJ/kg, $s_1 = 5.615$ kJ/kg K. For isentropic expansion from 1 to 2, $s_1 = s_2$; therefore $5.615 = 0.559 + 7.715 x_2$, so $x_2 = 0.655$. Hence $h_2 = 163 + 0.655(2409) = 1741$ kJ/kg, $h_3 = 163$ kJ/kg. Also $T_1 = 311$ °C $= 584$ K, $T_2 = 39$ °C $= 312$ K.

Carnot cycle efficiency $\eta_c = 1 - 312/584 = \mathbf{0.466}$

Rankine cycle efficiency $\eta_R = \dfrac{2725 - 1741}{2725 - 163} = \mathbf{0.384}$

SSC $= \dfrac{3600}{2725 - 1741} = \mathbf{3.66 \text{ kg/kWh}}$

In the Carnot cycle feed-pump work $= h_4 - h_3$. Now $s_4 = s_3$ (isentropic process); therefore $3.360 = 0.559 + x_3(7.715)$, and therefore $x_3 = 0.363$. Hence $h_3 = 163 + 0.363(2409) = 1038$ kJ/kg. The feed-pump work $= 1408 - 1038 = 370$ kJ/kg. The net work output is $(2725 - 1741) - 370 = 614$ kJ/kg, giving the SSC $= 3600/614 = \mathbf{5.86 \text{ kg/kWh}}$.

This shows the high reduction in net work output due to the high value of the feed-pump work.

In the Rankine cycle the feed-pump work is small, since the feed pump is handling a liquid, not a vapour. The specific volume of *water* is 0.0014 m³/kg at 100 bar, and 0.0010 m³/kg at 0.07 bar. The feed-pump work is approximately $v_f(p_4 - p_3) = 0.0012 \times 10^5(100 - 0.07)$ J/kg $= 12$ kJ/kg and can therefore be neglected.

In the second part of the problem the inlet steam is superheated. Referring to Fig. 3.7,

(a) $h_1 = 3097$, $s_1 = 6.213$, $s_{2s} = s$

therefore, $6.213 = 0.559 + 7.715 x_{2s}$

$x_{2s} = 0.733$

$h_{2s} = 163 + 0.733(2409)$

$= 1928$ kJ/kg

$\eta_c = 1 - 312/673 = \mathbf{0.536}$

$\eta_R = \dfrac{3097 - 1928}{3097 - 163} = \mathbf{0.398}$

SSC $= 3600/1169 = \mathbf{3.08 \text{ kg/kWh}}$.

(b) The turbine inefficiency is taken into account by the turbine isentropic efficiency

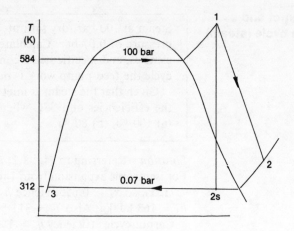

Figure 3.7

$$\frac{h_1 - h_2}{h_1 - h_{2s}} = 0.80$$

output is now reduced to 0.8(1169) = 935 kJ/kg, and $h_2 = 3097 - 935 = 2162$ kJ/kg.

$$\eta_R = \frac{935}{3097 - 163} = \mathbf{0.319} \text{ and SSC} = 3600/935$$
$$= \mathbf{3.85 \text{ kg/kWh}}.$$

It should be noted that the steam condition leaving the turbine is now at point 2. $h_2 = 2162 = 163 + 2409 x_2$, therefore $x_2 = 0.83$. The frictional effects in the turbine generate heat, which increases the enthalpy of the steam (reheat) and so increases the exhaust steam quality.

3.2 Reheat cycle

A turbine is supplied with steam at 100 bar, 400 °C. The condensor pressure is 0.07 bar. The steam is extracted at a point where the pressure is 40 bar, and reheated to 400 °C before completing the expansion. The isentropic efficiency of each stage = 80 %.
Determine the Rankine-cycle efficiency and specific steam consumption, by calculation or use of the h–s chart. Comment on the values obtained.

Solution Referring to Fig. 3.8, $h_1 = 3097$ kJ/kg, $s_1 = 6.213$.
In the first stage:

$$s_{2s} = s_1 = 6.213 \text{ at 40 bar}$$

Now $s_g = 6.070$, therefore steam is superheated to 274 °C, and $h_{2s} = 2879$ kJ/kg.

Isentropic enthalpy drop = $h_1 - h_{2s} = \Delta h_s = 218$ kJ/kg

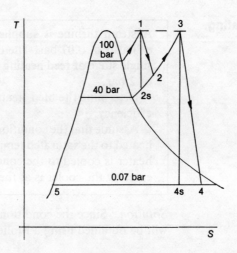

Figure 3.8

therefore

$$\text{actual drop} = \Delta h = 0.8(218) = 174 \text{ kJ/kg}.$$

Hence

$$h_2 = 3097 - 174 = 2923 \text{ kJ/kg}.$$

Considering the second stage: $s_{4s} = s_3$.

Therefore $6.769 = 0.559 + 7.715 x_{4s}$, so $x_{4s} = 0.805$. Hence

$$h_{4s} = 163 + 0.805(2409) = 2102 \text{ kJ/kg}$$

therefore

$$\Delta h_s = h_3 - h_{4s} = 1112 \text{ kJ/kg} \qquad \Delta h = 0.8 \Delta h_s = 890 \text{ kJ/kg}$$

so

$$h_4 = 3214 - 890 = 2324 \text{ kJ/kg}, \, x_4 = 0.90$$

Work output $= (h_1 - h_2) + (h_3 - h_4) = 1064$ kJ/kg

Heat input $= (h_1 - h_5) + (h_3 - h_2) = 3225$ kJ/kg

Therefore

$$\eta_R = 1064/3225 = \mathbf{0.330}$$

$$\text{SSC} = 3600/1064 = \mathbf{3.38 \text{ kg/kWh}}.$$

As obtained in the previous example 3.1(b) with a single stage of expansion and no reheat, $\eta_R = 0.319$ and the SSC = 3.85 kg/kWh. The effect of reheat therefore is to increase the Rankine cycle efficiency slightly (by 3.3 %), and reduce the SSC significantly (by 12.2 %). The main effect is to reduce the steam flow rate for a given output.

3.3 Feed heating cycle

A steam turbine is supplied with 10 kg/s at 20 bar, 300 °C and exhausts at 0.07 bar. The overall isentropic efficiency = 70 %. A single stage of feed heating is used with the steam bled at a pressure of 2 bar.

Determine the bled steam flow rate, power developed and cycle efficiency.

Assume that the condition curve is a straight line, the feed water is heated to the saturated temperature in the heater, the drain from the heater is cooled to the condensor temperature, and the feed water entering the cooler is at the condensor temperature.

Solution Since the condition line is assumed to be straight the solution will be obtained using a Molier ($h-s$) chart.

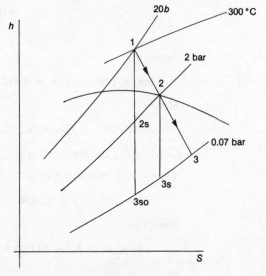

Figure 3.9

The values of enthalpy and entropy, determined from the chart, are shown in Table 3.1.

Table 3.1

Point	Condition	h(kJ/kg)	s(kJ/kg K)
1	20 bar, 300 °C	3025	6.77
3s0	0.07 bar	2070	6.77
3	0.07 bar, 91 % dry	2356	7.58
2	2 bar, dry	2710	7.13
2s	2 bar	2565	6.77
3s	0.07 bar	2220	7.13

The overall isentropic drop is first determined, giving point 3s0. The overall efficiency = 70 %, therefore $h_1 - h_3 = 0.7(h_1 - h_{3s0})$. The condition line can then be drawn, connecting points 1 and 3: its intersection with the 2 bar pressure line determines point 2.

An energy balance on the heater gives

$$mh_2 + h_{f4} = h_{f2} + mh_{f4}.$$

Therefore $2710m + 163 = 505 + 163m$; therefore $m = 0.134$ kg/kg boiler steam = **1.34 kg/s**.

The work done/kg boiler steam = $(3025 - 2710)$
$+ (1 - 0.134)(2710 - 2070) = 869$ kJ

therefore power output = **8690 kW**.

The heat input = $3025 - 505 = 2520$ kJ/kg; therefore cycle efficiency = $869/2520 = $ **0.345**.

(*Note:* Without the feed heating, the power output = $(3025 - 2070) = 955$ kJ, and the heat input = $3025 - 163 = 2862$ kJ, therefore cycle efficiency = $955/2862 = 0.334$. The effect of feed heating is therefore to increase the efficiency and reduce the power output/kg.)

Also the isentropic efficiency of each stage is

$$\frac{3025 - 2710}{3025 - 2565} = 0.69 \quad \text{and} \quad \frac{2710 - 2356}{2710 - 2220} = 0.72$$

Thus the stage efficiency is not equal to the overall efficiency.

3.4 Feed heating cycle

> A steam plant operates between a boiler pressure of 40 bar, dry saturated, and a condensor pressure of 0.03 bar.
>
> Calculate the Carnot and Rankine cycle efficiency, and specific steam consumptions.
>
> Steam is bled at 3.5 bar to a feed heater. Neglecting the feed-pump work, calculate the bled-steam rate, kg/kg boiler steam, Rankine cycle efficiency and specific steam consumption.

Solution Carnot cycle efficiency $\eta_c = 1 - T_{min}/T_{max} = 1 - 297/523 = $ **0.432**. Referring to Fig. 3.1 $s_2 = s_1 = 6.070 = 0.354 + 8.222x_2$, therefore $x_2 = 0.695$ and $h_2 = 101 + 2444(0.695) = 1800$ kJ/kg. Also $s_3 = s_4 = 2797 = 0.354 + 8.222x_3$; therefore $x_3 = 0.297$ and $h_3 = 101 + 0.297(2444) = 827$ kJ/kg.

Turbine work = $h_1 - h_2 = 2801 - 1800 = 1001$ kJ/kg.
Feed work = $h_4 - h_3 = 1087 - 827 = 260$ kJ/kg

therefore, net work = $1001 - 260 = 741$ kJ/kg. The net work could also be determined from the efficiency: net work = η_c (heat input) = $0.432(h_1 - h_4) = 741$ kJ/kg.
SSC = $3600/741 = $ **4.86 kg/kWh**.
Rankine-cycle efficiency

$$\eta_R = \frac{h_1 - h_2}{h_1 - h_3} \text{ (Fig. 3.2)} = \frac{1001}{2801 - 101} = \mathbf{0.371}$$

SSC = 3600/1001 = **3.60 kg/kWh**.

Assuming adiabatic mixing in the heater, and referring to Fig. 3.10,

$$mh_6 + (1 - m)h_3 = h_4$$

Now $s_1 = s_6 = 6.070 = 1.727 + 5.214x_6$ therefore $x_6 = 0.833$; therefore

$$h_6 = 584 + 0.833(2148) \text{ kJ/kg}$$
$$= 2373 \text{ kJ/kg}$$

Substituting,

$$2373m + 101(1 - m) = 584;$$

therefore $m = \mathbf{0.213 \text{ kg/kg}}$.

$$\text{The work output} = 1.0(h_1 - h_6) + (1 - 0.213)(h_6 - h_2)$$
$$= 879 \text{ kJ/kg boiler steam}$$

and the heat input = $h_1 - h_4 = 2217$ kJ/kg.
Therefore, cycle efficiency = 879/2217 = **0.396**
SSC = 3600/879 = **4.10 kg/kWh**.

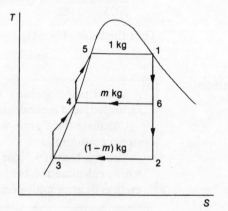

Figure 3.10

3.5 Binary vapour cycle

A binary vapour cycle uses mercury and water. The mercury is vapourised at 20 bar, and is dry saturated. It expands in a turbine to a pressure of 0.2 bar, and then passes to a heat exchanger. Turbine isentropic efficiency = 80 %.

The steam leaves the exchanger at 50 bar, dry saturated. It then passes through a superheater in which the temperature is raised to 500 °C. The steam then expands in a turbine, isentropic efficiency 80 %, to a condensor at a pressure of 0.05 bar. Calculate the kg mercury/kg steam, and cycle efficiency.

Briefly outline the reasons for using such a cycle.

Solution The flow and T–s diagrams are shown in Fig. 3.11. Considering the mercury cycle $h_1 = 366.5$ kJ/kg, $s_1 = 0.4890$ kJ/kg K.

$$s_{2s} = s_1 = 0.4890 = 0.0961 + 0.5334 x_{2s}$$

Therefore $x_{2s} = 0.7366$; therefore

$$h_{2s} = 38.05 + 0.7366(294.0) = 354.6 \text{ kJ/kg.}$$

The isentropic enthalpy drop is

$$\Delta h_s = 366.5 - 254.6 = 111.9 \text{ kJ/kg}$$

therefore $\Delta h = 0.8 \Delta h_s = 89.5$ and $h_2 = 277.0$ kJ/kg.

Considering the steam cycle $h_5 = 3433$ kJ/kg, $s_5 = 6.975$ kJ/kg K; therefore $s_{6s} = 6.975 = 0.476 + 7.918 x_{6s}$, so $x_{6s} = 0.821$. Hence

$$h_{6s} = 138 + 0.821(2423) = 2127 \text{ kJ/kg}, \qquad \Delta h_s = h_5 - h_{6s} = 1306$$

therefore $\Delta h = 0.8 \Delta h_s = 1045$ kJ/kg and $h_6 = 2388$ kJ/kg.

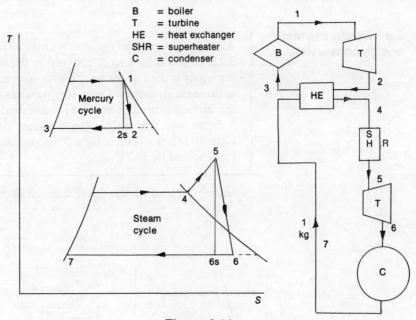

Figure 3.11

Let m = kg Hg/kg steam. An energy balance on the superheater gives $m(h_2 - h_3) = 1.0(h_4 - h_7)$.

Therefore, $m(277.0 - 38.05) = 2794 - 138$, therefore $m = $ **11.12 kg**.
Work output $= 1.0(h_5 - h_6) + m(h_1 - h_2) = 2040$ kJ/kg steam
Heat input $= m(h_1 - h_3) + 1.0(h_5 - h_4) = 4291$ kJ/kg steam
Therefore cycle efficiency $= 2040/4291 = $ **0.475**.

The critical point for steam is 211 bar, 374 °C. Thus to attain the maximum temperature set by metallurgical limits (approx. 600 °C) supercritical pressures or high superheat is required. The critical conditions for mercury are 75 bar, 753 °C so that by the use of mercury vapour the high temperature can be obtained at much lower working pressures than would be required by the use of steam.

However, the advantages of mercury do not apply to the bottom end of the cycle, at low temperature condensation. The low saturation pressure and high specific volume of the mercury vapour would require very large and expensive condensing equipment. Examples of the properties of steam and mercury are given in Table 3.2.

Table 3.2

	satn. temp. °C	specific volume (m³/kg)	p (bar)
steam	32.9	28.20	0.05
mercury	229.7	4.226	0.05
mercury	109.2	259.6	0.0006

3.6 Ocean thermal energy conversion

An ocean thermal energy conversion (OTEC) plant is shown in Fig. 3.12. In this solar energy conversion the top layer of warm surface sea water is used as the heat source to evaporate the working fluid, in this case ammonia. The colder water at a substantial depth below the surface is used as the sink, to condense the working fluid.

Determine the cycle efficiency and comment on the scheme. Calculate the cycle efficiency if the vapour leaving the evaporator is superheated by 50 °C.

Solution   At a saturation temperature of 24 °C, the pressure = 9.722 bar.

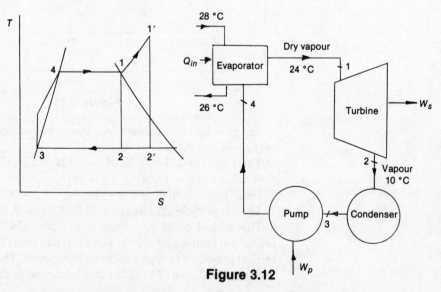

Figure 3.12

$h_1 = 1465.2$ kJ/kg, $h_3 = 294.1$ kJ/kg, since the saturation pressure corresponding to 10 °C is 6.149 bar.

$s_1 = 5.049 = 0.881 + 4.332x_2$; therefore

$$x_2 = 0.962 \quad \text{and} \quad h_2 = 227.8 + 0.962(1226.5) = 1407.7 \text{ kJ/kg}$$

Consider the system

Work output $= h_1 - h_2 = 57.5$ kJ/kg
Feed-pump work $= h_4 - h_3 = 66.3$ kJ/kg
Heat input $= h_1 - h_4 = 1171.1$ kJ/kg
Heat rejected $= h_2 - h_3 = 1179.9$ kJ/kg.

Total energy input $= Q_{in} + W_p = 1237.4$ kJ/kg, and the energy leaving the control volume $= W_s + Q_c = 1237.4$ kJ/kg. In this case therefore there is no net work output. This arrangement is then unsatisfactory.

If the vapour is superheated,

$$h_1 = 1602.7 \text{ kJ/kg} \quad s_1 = 5.478 = 0.881 + 4.332x_2$$

therefore $x_2 > 1$, i.e. vapour is still superheated. $T_2 = 55$ °C, $h_2 = 1535.4$ kJ/kg.

Therefore

$$\Delta h_s = 1602.7 - 1535.4 = 67.3 \text{ kJ/kg}$$
$$\text{net work} = 67.3 - 66.3 = 1 \text{ kJ/kg}$$

and

$$\text{cycle efficiency} = 1/(1602.7 - 227.8) = 0.0007.$$

The plant is now producing net power, but at a very low efficiency of 0.07 %. This emphasizes the practical difficulty of low-grade heat conversion into mechanical/electrical power! Hence although the source of heat is 'free', the cost of converting that heat is substantial.

3.7 Steam turbine characteristic

A 10 MW steam turbogenerator has the following characteristic

$$I(\text{MW}) = 3.4 + 2.5L + 0.01L^2$$

where I = heat input, L = load (MW).

The capital cost = £120/kW; fuel cost = 0.1 p/MJ; labour, maintenance costs = £30 000 (1 + CF) p.a.; and the fixed charge rate is 12 %.

Plot a graph of thermal efficiency and unit cost (p/kWh) against the capacity factor CF, based on a year of 6000 h. Comment on the curves.

Solution The capacity factor CF = mean load/rated capacity. The thermal efficiency = L/I.

Considering the total annual cost, the total is made up of the following items

(a) fixed cost = 12 % of the capital cost
    = 0.12 × 120 × 10 000 = £144 000
(b) labour, maintenance cost = £30 000 (1 + CF)
(c) fuel cost = I × 6000 × 3600 × 0.1 p = £21 600I

The unit cost (cost per unit of electricity generated) is

$$\frac{C \times 100 \times 3.6}{L \times 6000 \times 3600} = \frac{5}{3} \times 10^{-5} C/L$$

where C = total annual cost, £.

It should be noted that 1 unit of electricity, 1 kWh = 3.6 MJ.

The values of thermal efficiency and unit cost are calculated for a range of values of CF and shown in Table 3.3.

Table 3.3

CF (%)	0	10	20	30	40	50	60	70	80	90	100
L (MW)	0	1	2	3	4	5	6	7	8	9	10
I (MW)	3.4	5.91	8.44	10.99	13.56	16.15	18.76	21.39	24.04	26.71	29.40
(%)	—	16.92	23.70	27.30	29.50	30.96	32.0	32.72	33.28	33.70	34.01
Cost: fixed	144	144	144	144	144	144	144	144	144	144	144
labour	30	33	36	39	42	45	48	51	54	57	60
fuel	—	127.6	182.3	237.4	292.9	348.8	405.2	462.0	519.3	576.9	635.0
total (£p.a.)	174	304.6	362.3	420.4	478.9	537.8	597.2	657.0	717.3	777.9	839
Unit cost (p/kWh)	—	5.077	3.019	2.335	1.995	1.793	1.659	1.564	1.494	1.441	1.398

The resulting graphs are shown in Fig. 3.13. The efficiency rises rapidly at low capacity factors, and flattens off at higher factors. The efficiency lies in the range 30–35 % over a capacity factor range of 42–100 %.

The unit cost of generation is high at low capacity factors, becoming almost constant at capacity factors > 50 %. The fuel cost is the major item, and this can be seen from Table 3.4, showing the fuel costs as a % of the total cost.

Table 3.4

CF (%)	10	20	30	40	50	60	70	80	90	100
fuel cost (%)	41.9	50.3	56.5	61.2	64.9	67.8	70.3	72.4	74.2	75.7

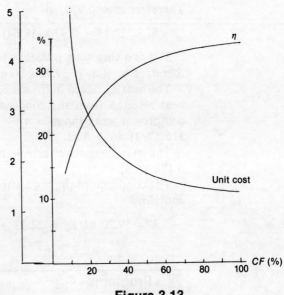

Figure 3.13

3.8 Refrigeration cycle

A vapour compression refrigerator runs between temperature limits of −15 °C and +25 °C, using Freon-12 as the working fluid. The refrigeration effect = 20 kW. Calculate the coefficient of performance and flow rate of refrigerant; and condition of the refrigerant after throttling.

Solution Referring to Fig. 3.14, $h_2 = 197.73$ kJ/kg and

$$s_2 = 0.6869 = 0.0906 + 0.6145 x_1$$

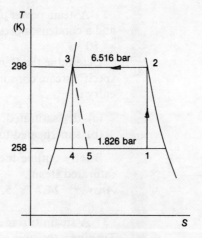

Figure 3.14

VAPOUR CYCLES 47

Therefore $x_1 = 0.970$, so

$$h_1 = 22.33 + 0.970(158.64) = 176.27 \text{ kJ/kg}$$

3–4 is a throttling process which is adiabatic but not reversible, and therefore $h_3 = h_5 = 59.70$ kJ/kg.

The heat extracted in the evaporator $= h_1 - h_5 = 116.57$ kJ/kg. The heat rejected from the condenser $= h_2 - h_3 = 138.03$ kJ/kg. The compressor work input $= h_2 - h_1 = 21.46$ kJ/kg. Hence the COP $= 116.57/21.46 = $ **5.43**.

The refrigeration effect $= m(116.57) = 20$ kW, therefore $m = $ 0.172 kg/s.

The condition of the fluid after throttling can be determined from the enthalpy:

$$h = 59.70 \text{ kJ/kg} = 22.33 + 158.64 x_5, \text{ therefore } x_5 = \textbf{0.236}.$$

Problems

1 A steam turbine is supplied with steam at a pressure of 100 bar, and exhausts to a condensor at a pressure of 0.10 bar. Turbine isentropic efficiency = 70 %.

Determine the Rankine cycle efficiency and steam condition at the turbine exit, and specific steam consumption when the initial steam condition is

(a) superheated to 350 °C,
(b) superheated to 400 °C.

Answer   (a) 26.7 %, 0.837 dry, 4.93 kg/kWh; (b) 27.2 %, 0.844 dry, 4.55 kg/kWh

2 A steam power plant operates between a boiler pressure of 40 bar and a condensor pressure of 0.25 bar. Turbine isentropic efficiency = 80 %.

Determine the Rankine cycle efficiency, using the h–s chart, and specific steam consumption (kg/kWh), when the steam at the boiler entry is

(a) dry saturated,
(b) superheated to 300 °C.

Briefly outline the advantages of using superheated steam over saturated steam.

Answer   24.7 %, 5.77:25.0, 5.34

3 A steam turbine exhausts to a condensor pressure of 0.3 bar. Turbine isentropic efficiency = 75 %. Determine the Rankine and Carnot cycle efficiencies, and specific steam consumption when the boiler steam is dry saturated, at 30, 60, 90, 120, 150, 180 bar. The $h-s$ chart should be used.

Plot the Rankine cycle efficiency and SSC against boiler pressure, and comment on the graph.

4 Steam is supplied at 50 bar, 350 °C to a turbine and the condensor pressure is 0.04 bar. Turbine isentropic efficiency = 80 %.

Calculate /kg steam:

(a) the turbine work output,
(b) the feed-pump work,
(c) the heat transferred to the cooling water,
(d) the heat supplied,
(e) the Rankine cycle efficiency,
(f) the specific steam consumption.

Answer 902, 5, 2047, 2954 kJ/kg; 30.4 %; 3.99 kg/kWh

5 Steam is supplied to a turbine at 50 bar, 400 °C. It expands in the first stage until it is dry saturated, and then reheated to 400 °C before expanding in the second stage to a condensor pressure of 0.05 bar.

Calculate the work output, SSC and Rankine cycle efficiency.

Compare these values with those obtained by a single-stage expansion without reheat.

Answer 1390 kJ/kg, 2.59 kg/kWh, 39.0 %; 1170, 3.08, 38.3 %

6 A steam turbine is required to give a power output of 300 MW. The boiler steam is supplied at 150 bar, 400 °C. The steam expands in the first stage (isentropic efficiency 80 %) to a pressure of 40 bar, at which 2 % of the steam is extracted for feed heating. The remainder is then expanded in the second stage (isentropic efficiency 75 %) to a condensor pressure of 0.05 bar.

Determine the boiler output, tonnes/h.

Answer 1193

7 A turbine operates on a regenerative cycle. Steam is supplied at 35 bar, dry saturated and exhausted at 0.06 bar. The condensate is pumped to a pressure of 3 bar at which it is mixed with steam bled from the turbine at 3 bar.

The mixture (which is at the saturation temperature) is then pumped to the boiler.

Neglecting the feed-pump work, calculate the kg bled steam per kg boiler steam; specific steam consumption; and ideal cycle efficiency.

Answer 0.184 kg/kg; 4.35 kg/kWh; 36.9 %

8 The pressure in the evaporator of a refrigerator, using ammonia, is 1.902 bar and in the condensor 12.37 bar. Calculate the ideal *COP* when working between the saturation temperatures, and the ideal refrigerating effect.

In a real cycle between the same pressures the dry saturated vapour is isentropically compressed and the vapour throttled between the condensor and compressor. Calculate the *COP* and refrigerating effect, and mass flow rate of ammonia, per tonne of refrigeration effect.

Answer   4.87, 943 kJ/kg; 3.90, 1087 kJ/kg, 11.64 kg/h

9   A Rankine cycle uses steam as the working fluid. Steam leaves the boiler and enters a turbine at 100 bar, 500 °C. After expansion in the turbine to 5 bar the steam is reheated to 500 °C and then expanded in a low pressure turbine to 0.1 bar.

Determine the cycle efficiency, work ratio and specific steam consumption. Sketch the cycle on $T-s$ and $h-s$ diagrams.

Briefly explain the advantages and disadvantages of using a reheat process in a Rankine cycle.

Answer   0.409; 2.20 kg/kWh

10   Steam leaves a boiler 98 % dry, at 100 bar, and passes through a superheater where its temperature is raised to 500 °C before being admitted to a steam turbine, having an isentropic efficiency of 75 %.

The steam leaves the turbine at 0.2 bar and is returned to the boiler via a condensor, hot well and feed pump. Neglecting changes in potential and kinetic energy throughout the system determine the feed-pump work/kg steam generated; the Rankine and Carnot efficiency for the cycle.

Answer   4.4 kJ; 0.289, 0.569

11   Steam is delivered to a two-stage turbine at 50 bar, 450 °C. It is expanded in the first stage to the dry saturated condition, reheated to 450 °C, and then expanded in the second stage to 0.18 bar. Given that both stages of the turbine have an isentropic efficiency of 82 %, determine

(a) the pressure and enthalpy of the steam leaving the first stage;

(b) the enthalpy of the steam entering and leaving the second stage;

(c) the thermal efficiency of the plant, including the feed-pump work, and assuming the steam leaves the condensor as saturated liquid.

Answer   (a) 1.8 bar,   2702 kJ/kg,   (b) 3382   and   2812 kJ/kg,   (c) 0.315

4

Internal combustion engines

The conversion of chemical energy in a fuel into thermal energy (heat), and then mechanical energy is one of the important areas of energy conversion. In this chapter the conversion in internal combustion (I.C.) engines is considered, where the mechanical energy is in the form of reciprocating machinery. The use of petrol, diesel, gas engines is extensive in the areas of transport and electricity generation.

Air standard cycles

The thermodynamics of the I.C. engine cannot strictly be considered as a series of processes comprising a cycle, but rather a set of unsteady flow processes. However, as a basis of comparison the air standard cycle is used, in which it is assumed that the working fluid throughout the thermodynamic cycle is air, behaving as an ideal gas. The three cycles used are shown in Figs 4.1–3 and comprise the constant-volume (Otto) cycle, the constant-pressure (Diesel) cycle, and the dual-combustion cycle.

The thermal efficiency of the cycle, known as the air standard efficiency ASE, is

$$\text{Constant-volume cycle } ASE = 1 - \left(\frac{1}{CR}\right)^{\gamma-1}$$

$$\text{Constant-pressure cycle } ASE = 1 - \frac{\alpha^{\gamma} - 1}{\gamma(\alpha - 1)}\left(\frac{1}{CR}\right)^{\gamma-1}$$

$$\text{Dual-combustion cycle } ASE = 1 - \frac{\alpha^{\gamma}\beta - 1}{\gamma\beta(\alpha - 1) + (\beta - 1)}\left(\frac{1}{CR}\right)^{\gamma-1}$$

where CR = compression ratio = v_1/v_2
α = cut-off ratio = v_3/v_2 (Fig. 4.2) or v_4/v_3 (Fig. 4.3)
β = pressure ratio = p_3/p_2 (Fig. 4.3)

The cycles are so named because the heat supplied is respectively at constant volume, constant pressure, and at both.

The net work output is represented on the indicator (pv) diagram by the area enclosed by the process curves, and can be expressed in terms of the indicated mean effective pressure

$IMEP$ = net work output/swept volume

It should be noted that in these cycles the heat exchanges Q_{in} and Q_{rej} are energy transfers across the system boundary.

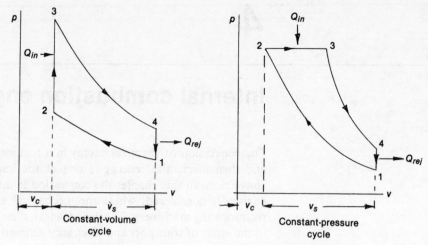

Figure 4.1 — Constant-volume cycle

Figure 4.2 — Constant-pressure cycle

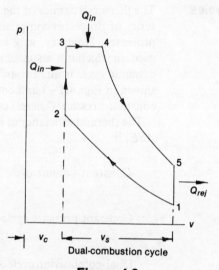

Figure 4.3 — Dual-combustion cycle

Indicator diagrams: characteristics

In a real engine the indicator diagram differs from the ideal air standard diagram for several reasons: the thermal efficiency is also less than the *ASE*. The working fluid in a real engine is not air throughout the cycle, but an air–fuel mixture – combustion products. The properties of the fluid therefore vary. The fluid does not behave as an ideal gas, and the specific heat variation with temperature can be of importance. Further differences from the ideal occur because the heat supply and rejection are not instantaneous (i.e. at constant volume), and the compression and expansion processes are neither reversible (frictional effects) nor adiabatic (cooling of the cylinder): a temperature drop between the fluid and cylinder wall occurs.

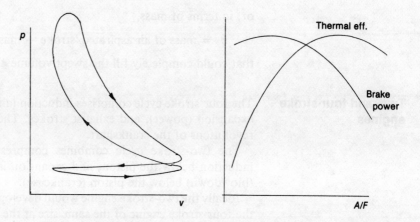

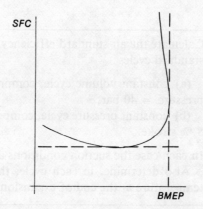

Figure 4.4

A typical indicator diagram and characteristics are shown in Fig. 4.4. The thermal efficiency is based upon the power developed in the cylinder (indicated), or that taken off at the shaft (brake). The difference between them is a measure of the mechanical losses between the cylinder and shaft, and can be substantial.

Similarly the *MEP* can be based upon the brake (shaft) power: *BMEP* = brake power/swept volume.

The efficiency can also be expressed in terms of the specific fuel consumption *SFC* = kg fuel/kWh output.

The volume of air drawn into the cylinder is, in practice, less than the swept volume. A clearance volume is necessary so that the residual gases in the cylinder expand to a greater volume on the suction stroke.

This effect is measured by the volumetric efficiency, η_v, where

$$\eta_v = \text{volume of aspirated air at STP/swept volume}$$

Two- and four-stroke engines

or, in terms of mass,

η_v = mass of air aspirated/stroke ÷ mass of air, at STP,

that could completely fill the swept volume.

The four-stroke cycle comprises induction (air or air/fuel), compression, expansion (power), and exhaust strokes. The cycle is completed in two revolutions of the crankshaft.

The two-stroke cycle combines compression in the cylinder with induction below the piston; and expansion in the cylinder with exhaust (blowdown) below the piston (crankcase).

Ideally the two-stroke engine would develop a larger power output than the four-stroke engine of the same size at the same speed. In practice the efficiency of the two-stroke engine is less than that of the four-stroke engine, though it can be improved by supercharging.

4.1 Air standard cycles

> Calculate the air standard efficiency and *MEP* of the following air standard cycles
>
> (a) constant volume cycle, compression ratio = 9:1, maximum pressure = 40 bar,
> (b) constant pressure cycle, compression ratio = 18:1, cut-off at 5 % stroke.
>
> In each case the suction conditions are 1 bar, 15 °C.
> Also determine, in each cycle, the maximum temperature and temperature at the end of expansion.

Solution The cycles are shown in Fig. 4.5, where for convenience v_2 is taken as 1.0 m³.

(a) Processes 1–2 and 3–4 are isentropic; therefore pv^γ = constant. The air is taken as an ideal gas, therefore $pv = mRT$ and the specific heat is constant.

Thus, $T_2 = T_1(v_1/v_2)^{\gamma-1} = 288(9)^{0.4} = 694$ K

$p_2 = p_1(v_1/v_2)^\gamma = 1.0(9)^{1.4} = 21.67$ bar

Process 2–3 is at constant volume; therefore

$$\frac{T_3}{T_2} = \frac{p_3 v_3}{p_2 v_2} = \frac{p_3}{p_2}$$

Therefore $T_3 = 694(40/21.67) = 1281$ K

$$p_4 = p_3(v_3/v_4)^\gamma = 40\left(\frac{1}{9}\right)^{1.4} = 1.85 \text{ bar}$$

$$T_4 = T_3(v_3/v_4)^{\gamma-1} = \left(\frac{1}{9}\right)^{0.4} \times 1281 = 532 \text{ K}$$

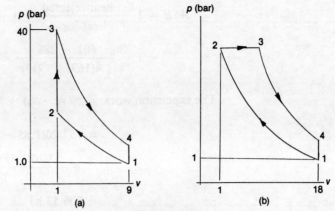

Figure 4.5

$$ASE = \frac{\text{heat rejected}}{\text{heat input}} = 1 - \frac{C_v(T_4 - T_1)}{C_v(T_3 - T_2)}$$

$$= 1 - \frac{532 - 288}{1281 - 694} = \mathbf{0.585}$$

or it can be calculated from $ASE = 1 - \left(\dfrac{1}{CR}\right) = \left(\dfrac{1}{9}\right)^{0.4} = \mathbf{0.585}$.

$IMEP$ = area of loop/swept volume.

The expansion work

$$\int_3^4 p\, dv = \frac{p_3 v_3 - p_4 v_4}{\gamma - 1} = \frac{(40 - 1.85 \times 9) \times 1.0 \times 10^5}{0.4}$$

$$= 58\,380\ \text{J}$$

and the compression work

$$\int_1^2 p\, dv = \frac{p_2 v_2 - p_1 v_1}{\gamma - 1} = \frac{(21.67 - 1.0 \times 9) \times 1.0 \times 10^5}{0.4}$$

$$= 31\,680\ \text{J}$$

Therefore net work = 58.38 − 31.68 = 26.70 kJ
Therefore $IMEP$ = 26.70/(9 − 1) = **3.34 bar** or **334 kN/m²**.

(b) The heat input is cut off at 5% of the stroke, i.e. $v_3 - v_2 = 0.05(v_1 - v_2) = 0.05(17) = 0.85$, therefore $v_3 = 1.85$ m³. Again processes 1–2 and 3–4 are isentropic.

$T_2 = T_1(v_1/v_2)^{\gamma-1} = 288(18)^{0.4} = 915$ K
$p_2 = p_1(v_1/v_2)^{\gamma} = 1.0(18)^{1.4} = 57.20$ bar
$T_3 = T_2(pv_3/v_2) = 915(1.85) = 1693$ K
$T_4 = T_3(v_3/v_4)^{\gamma-1} = 1694(1.85/18)^{0.4} = 681$ K
$p_4 = p_3(v_3/v_4)^{\gamma} = 2.37$ bar

$$ASE = 1 - \frac{\text{heat rejected}}{\text{heat input}} = 1 - \frac{C_v(T_4 - T_1)}{C_p(T_3 - T_2)}$$

$$= 1 - \frac{681 - 288}{1.4(1693 - 915)} = \mathbf{0.639}$$

The expansion work $= p_2(v_3 - v_2) + \dfrac{p_3 v_3 - p_4 v_4}{\gamma - 1}$

$$= \{57.20(1.85 - 1)$$
$$+ \frac{57.20(1.85) - 2.37(18)}{0.4}\} \times 10^5 \times 10^{-3}$$

$$= 206.52 \text{ kJ}$$

and the compression work $= \dfrac{p_2 v_2 - p_1 v_1}{\gamma - 1}$

$$= \frac{57.20 - 1.0(18)}{0.4} \times 10^5 \times 10^{-3}$$

$$= 98.0 \text{ kJ}$$

The net work is $206.52 - 98.0 = 108.52$ kJ, giving

$$IMEP = 108.52/17 = \mathbf{6.38 \text{ bar}} \text{ or } \mathbf{638 \text{ kN/m}^2}.$$

4.2 Air standard cycles

Derive an expression for the air standard efficiency of a dual combustion cycle in terms of the compression ratio, cut-off ratio and pressure ratio.

In such a cycle the suction conditions are 1 bar, 15 °C; compression ratio = 18:1; cut-off is at 5 % of the stroke, and the maximum pressure is 70 bar. Determine the ASE, IMEP, and maximum temperature in the cycle.

Solution The cycle is shown in Fig. 4.6.

Let $CR = v_1/v_2$, $\alpha = v_4/v_3$, $\beta = p_3/p_2$.

$T_2 = T_1(CR)^{\gamma-1}$, $p_2 = p_1(CR)^\gamma$

$T_3 = T_2 \dfrac{p_3 v_3}{p_2 v_2} = T_1(CR)^{\gamma-1}\beta$, $p_3 = \beta p_2$

$T_4 = T_3 \dfrac{p_4 v_4}{p_3 v_3} = \alpha\beta(CR)^{\gamma-1}T_1$

$T_5 = T_4(v_5/v_4)^{\gamma-1} = \alpha\beta(CR)^{\gamma-1}T_1(\alpha/CR)^{\gamma-1}$
$\quad = \alpha^\gamma \beta T_1$

$p_5 = p_4(v_4/v_5)^\gamma = p_3(\alpha/CR)^\gamma = \beta\alpha^\gamma p_1$

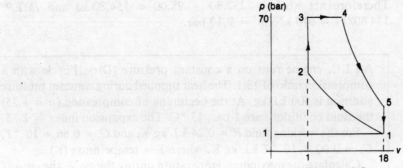

Figure 4.6

$$ASE = 1 - \frac{C_v(T_5 - T_1)}{C_v(T_3 - T_2) + C_p(T_4 - T_3)}$$

$$= 1 - \frac{T_5 - T_1}{T_3 - T_2 + \gamma(T_4 - T_3)}$$

$$= 1 - \frac{\alpha^\gamma \beta - 1}{\beta(CR)^{\gamma-1} - (CR)^{\gamma-1} + \gamma[\alpha\beta(CR)^{\gamma-1} - \beta(CR)^{\gamma-1}]}$$

$$= 1 - \frac{\alpha^\gamma \beta - 1}{(\beta - 1) + \gamma\beta(\alpha - 1)} \left(\frac{1}{CR}\right)^{\gamma-1}$$

In this problem $T_1 = 288$ K, $p_1 = 1$ bar. Also

$$v_4 - v_3 = 0.05(v_1 - v_2) = 0.85$$

Therefore $v_4 = 1.85$ m^3 (taking $v_2 = 1.0$ m^3)

$T_2 = 288(18)^{0.4} = 915$ K, $p_2 = 1.0(18)^{1.4} = 57.20$ bar
$T_3 = 915(70/57.20) = 1120$ K
$T_4 = 2072(1.85/18)^{0.4} = 834$ K, $p_5 = 70(1.85/18)^{1.4} = 2.90$ bar

Now $\alpha = 1.85$, $\beta = 70/57.20 = 1.224$, $CR = 18$.

Therefore, $ASE = 1 - \dfrac{(1.85)^{1.4}(1.224) - 1}{0.224 + 1.4(1.224)(0.85)} \left(\dfrac{1}{18}\right)^{0.4} = \mathbf{0.645}$

The expansion work $= p_3(v_4 - v_3) + \dfrac{p_4 v_4 - p_5 v_5}{\gamma - 1}$

$$= \left\{70(0.85) + \frac{70(1.85) - 2.90(18)}{0.4}\right\} \times 10^5 \times 10^{-3}$$

$$= 252.80 \text{ kJ}$$

and the compression work $= \dfrac{p_2 v_2 - p_1 v_1}{\gamma - 1}$

$$= \left\{\frac{57.20 - 1.0(18)}{0.4}\right\} \times 10^5 \times 10^{-3}$$

$$= 98.00 \text{ kJ}$$

INTERNAL COMBUSTION ENGINES 57

Therefore net work = 252.80 − 98.00 = 154.80 kJ and *IMEP* = 154 80/17 = 910 kN/m² = **9.10 bar**.

4.3 Variation of specific heat with temperature

> An I.C. engine runs on a constant pressure (Diesel) cycle with a compression ratio of 18:1. The heat supplied during constant pressure addition is 900 kJ/kg. At the beginning of compression ($n = 1.35$) the fluid conditions are 1 bar, 13 °C. The expansion index = 1.25.
> For the working fluid $R = 0.24$ kJ/kg K, and $C_v = 0.66 + 10^{-4}T$, $C_p = 0.90 + 10^{-4}T$ kJ/kg K, where T = temperature (K).
> Calculate the maximum temperature during the cycle, the cut-off ratio, and thermal efficiency.

Solution In this problem the cycle is nearer to a real situation in that the specific heat varies with the temperature, and the compression and expansion processes are not isentropic.

Referring to Fig. 4.2

$p_2 v_2 = p_1 v_1^n$, therefore $p_2 = 1.0(18)^{1.35} = 49.50$ bar

$T_2 = T_1 \dfrac{p_2 v_2}{p_1 v_1}$, therefore $T_2 = 286 \left(\dfrac{49.50}{1.0}\right)\left(\dfrac{1}{18}\right) = 798$ K

$$W = \int_1^2 p\,dv = \dfrac{p_2 v_2 - p_1 v_1}{n - 1} = \dfrac{10^5}{0.35}(49.50 - 18)v_2$$

$$= 9 \times 10^6 v_2 \text{ J/kg} = 9000 v_2 \text{ kJ/kg}$$

The compression is a non-flow process: therefore the energy equation becomes

$$Q - W = \int_1^2 C_v\,dT = \int_{286}^{798} (0.66 + 10^{-4}T)\,dT$$

$$= (0.66T + \tfrac{1}{2} \times 10^{-4}T) = 365.8 \text{ kJ/kg}$$

The work done = $9000 v_2$ kJ/kg, where $v_2 = v_1/18$, and $v_1 = mRT_1/p_1 = 1.0(240)(286)/(1 \times 10^5) = 0.686$ m³/kg.

Therefore $W = -9000(0.686)/18 = -343.2$ kJ/kg (compression work is negative).

Therefore $Q = -343.2 + 365.8 = 22.6$ kJ/kg.

Considering the heat addition at constant pressure,

$$W = \int_2^3 p\,dv = p_2(v_3 - v_2) = p_3(\alpha - 1)v_2, \text{ and}$$

$$Q = \int_2^3 C_p\,dT = \int_{798}^{T_3} (0.90 + 10^{-4}T)\,dT$$

$$= (0.90T + 0.5 \times 10^{-4}T^2)_{798}^{T_3}$$

$$= 0.9(T_3 - 798 + 0.5 \times 10^{-4}T_3^2 - 636\,804) \text{ kJ/kg}$$

The heat supplied is given as 900 kJ/kg.

Therefore $900 = 0.9(T_3 - 798) + 0.5 \times 10^{-4}(T_3^2 - 636\,804)$, therefore $T_3 = \mathbf{1677\ K}$.

The cut-off ratio can now be determined

$$\frac{T_3}{T_2} = \frac{p_3 v_3}{p_2 v_2}$$

Therefore $1677/798 = \alpha = \mathbf{2.10}$.

The work done is

$$p_3(\alpha - 1)v_2 = 49.50 \times 10^5 \times 1.10 \times 0.686/18 \text{ J/kg} = 207.5 \text{ kJ/kg}.$$

For the expansion process $n = 1.25$. Therefore

$$p_4 = p_3(v_3/v_4)^n = 49.50(2.10/18)^{1.25} = 3.38 \text{ bar},$$

and

$$T_4 = T_3(p_4 v_4 / p_3 v_3) = 1677 \left(\frac{3.38}{49.50}\right)\left(\frac{18}{2.10}\right) = 982 \text{ K}$$

The expansion work

$$= \frac{p_3 v_3 - p_4 v_4}{n - 1} = \frac{10^5}{0.25}(49.50 \times 2.10 v_2 - 3.38 \times 18 v_2)$$

$$= 17240 v_2 \text{ kJ/kg} = 657.0 \text{ kJ/kg}$$

$$Q - W = \int_3^4 C_v \, dT = \int_{982}^{1677} (0.66 + 10^{-4} T) \, dT$$

$$= -551.1 \text{ kJ/kg}$$

Therefore, $Q = +657.0 - 551.1 = 105.9 \text{ kJ/kg}$.

The net work done $= 207.5 + 657.0 - 343.2 = 521.3 \text{ kJ/kg}$, therefore thermal efficiency $= 521.3/900 = \mathbf{0.579}$.

It is of interest to compare the thermal efficiency with the

$$ASE = 1 - \frac{\alpha^\gamma - 1}{\gamma(\alpha - 1)}\left(\frac{1}{CR}\right)^{\gamma - 1}$$

$$= 1 - \frac{(2.1)^{1.4} - 1}{1.4(1.1)}\left(\frac{1}{18}\right)^{0.4} = \mathbf{0.627}$$

4.4 Engine parameters

A four-cylinder four-stroke engine develops 88 kW brake power at a speed of 4000 rev/min. The suction pressure is 0.95 bar, and temperature = 15 °C. The maximum pressure = 60 bar. Compression ratio = 9:1. The brake specific fuel consumption = 0.3 kg/kWh. The calorific value of the fuel = 42 MJ/kg. The indicated mean effective pressure = 60 % of the air standard cycle value. Given that the bore = stroke, calculate the mechanical efficiency, cylinder bore, mean piston speed, and air:fuel ratio. The volumetric efficiency = 85 %.

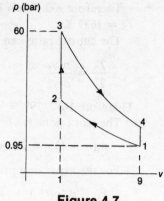

Figure 4.7

Solution Referring to the air standard cycle (Fig. 4.7),

$p_2 = 0.95(9)^{1.4} = 20.59$ bar

$T_2 = 288\left(\dfrac{20.59}{0.95}\right)\left(\dfrac{1}{9}\right) = 693$ K

$T_3 = \dfrac{60 \times 693}{20.59} = 2019$ K

$p_4 = 60(1/9)^{1.4} = 2.77$ bar

$T_4 = 2019(2.77/50)(9) = 839$ K

$IMEP = \left(\dfrac{p_3 v_3 - p_4 v_4}{\gamma - 1} - \dfrac{p_2 v_2 - p_1 v_1}{\gamma - 1}\right)/(v_1 - v_2) = 7.20$ bar

Therefore actual $IMEP = 0.6(7.2) = 4.32$ bar.

The fuel consumption $= 0.3 \times 88 = 26.4$ kg/h.

Therefore energy input $= \dfrac{26.4}{3600} \times 42\,000 = 308$ kW.

The $ASE = 1 - \left(\dfrac{1}{9}\right)^{0.4} = 0.585$, therefore indicated thermal efficiency = $0.6(0.585) = 0.351$, therefore indicated power output $= 0.351 \times 308 = 108$ kW. Hence the mechanical efficiency $= 88/108 = \mathbf{0.815}$.

The indicated power $= IMEP \times LAS$ where L = stroke, A = cylinder area, S = number of working strokes/s $= \tfrac{1}{2}N$ for a 4-stroke cycle. Substituting

$\tfrac{1}{4} \times 108$ kW $= 4.32 \times 10^5 (\text{N/m}^2) \times L(\text{m}) \times \dfrac{\pi D^2}{4}(\text{m}^2) \times$

$\tfrac{1}{2}\left(\dfrac{4000}{60}\right)(\text{m/s}) \times 10^{-3}$

therefore, $LD^2 = 0.0075$ and since $L = D$, $D = 0.196$ m or **196 mm** (each cylinder).

The mean piston speed = stroke/time taken for 1 stroke =

$$\frac{0.196 \times 4000}{120} = 6.53 \text{ m/s}$$

The swept volume = $\frac{\pi D^2}{4} L = 0.0059 \text{ m}^3$ and since the volumetric efficiency = 85 %, the volume of air induced/cylinder = 0.85(0.0059) = 0.00502 m³. At the suction conditions of 0.95 bar, 15 °C the air density is

$$\rho = p/RT = \frac{0.95 \times 10^5}{287 \times 288} = 1.149 \text{ kg/m}^3$$

therefore air induced/cylinder = $0.00502 \times 1.149 = 0.00577$ kg.

The fuel used/stroke

$$\frac{26.4}{60} \times \frac{1}{4000 \times 2} = 5.5 \times 10^{-5} \text{ kg}$$

Hence the air:fuel ratio is

$$\frac{0.00577}{5.5 \times 10^{-5} \times 4} = 26.2 \text{ kg/kg}$$

4.5 Engine test

A 4-cylinder, 4-stroke I.C. engine gave the results shown in Table 4.1 from a test.

Table 4.1

Speed N (rev/min)	Full load		Part load	
	Brake power P (kW)	Fuel consumption m_f (kg/s)	Brake power P (kW)	Fuel consumption m_f (kg/s)
5000	41.0	0.003 7	20.9	0.002 1
4500	38.8	0.003 4	19.4	0.001 8
4000	34.3	0.002 9	17.2	0.001 6
3500	31.0	0.002 5	15.7	0.001 4
3000	27.7	0.002 2	13.4	0.001 1
2500	23.1	0.001 9	11.6	0.001 0
2000	17.6	0.001 5	9.9	0.000 85
1500	11.6	0.001 1	6.0	0.000 55
1000	6.7	0.000 7	3.4	0.000 4

Construct graphs of brake power and specific fuel consumption plotted against the speed, and comment on the curves obtained.

Solution The $SFC = m_f \times 3600/P$ kg/kWh. The calculated values of the *SFC* are shown in Table 4.2.

Table 4.2

N (rev/min)	5000	4500	4000	3500	3000	2500	2000	1500	1000
Full load *SFC*	0.325	0.315	0.304	0.290	0.286	0.296	0.307	0.342	0.376
Part load *SFC*	0.361	0.334	0.335	0.321	0.296	0.310	0.309	0.330	0.423

The curves are shown in Fig. 4.8, and show that the part load consumption is significantly greater than that at full load. The difference decreases at low and high speeds.

4.6 Engine test

A 4-cylinder, 4-stroke petrol engine runs at a constant speed of 3000 rev/min. Cylinder bore = 75 mm, stroke = 100 mm.

A test at a constant throttle opening and variable mixture strength gave the results shown in Table 4.3.

Table 4.3

Brake power (kW)	31.3	31.8	31.5	31.3	30.6	29.6
Fuel consumption (kg/s)	0.0033	0.0031	0.0028	0.0026	0.0024	0.0023
Air:fuel ratio (kg/kg)	10.6	11.5	12.0	13.9	15.2	16.3
	28.0	26.5				
	0.0021	0.0020				
	17.3	18.0				

Plot graphs of brake mean effective pressure and specific fuel consumption against the air:fuel ratio.

If the fuel analysis (gravimetric) is 83 % C, 16 % H_2 calculate the stoichiometric air:fuel ratio.

Comment on the curves obtained.

Solution The $SFC = m_f \times 3600/P$ where m_f = fuel consumption (kg/s), and P = brake power (kW). Also

$$P = BMEP \text{ (bar)} \times 10^5 \times LA \times S \times n \text{ (W)}$$

where L = cylinder stroke (m), A = cylinder area (m²), S = number of working strokes/s, n = number of cylinders.

Therefore

$$P = BMEP \times 10^5 \times \frac{100}{1000} \times \frac{\pi}{4}(0.075)^2 \times \frac{3000}{600} \times \tfrac{1}{2} \times 4 \times 10^{-3} \text{ kW}$$

$$BMEP = 0.226P \text{ (bar)}$$

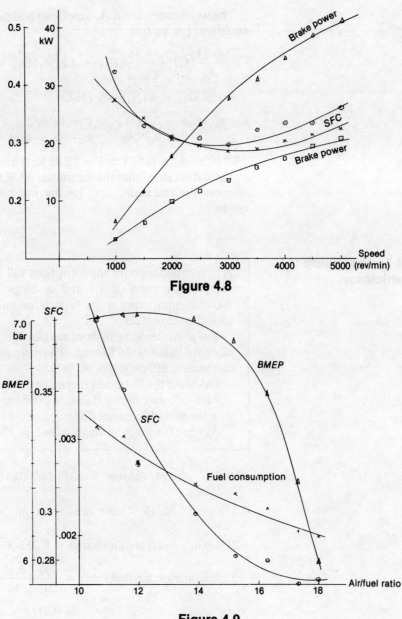

Figure 4.8

Figure 4.9

Table 4.4

A/F ratio	10.6	11.5	12.0	13.9	15.2	16.3	17.3	18.0
BMEP (bar)	7.074	7.187	7.119	7.074	6.916	6.690	6.328	5.989
SFC (kg/kWh)	0.380	0.351	0.320	0.299	0.282	0.280	0.270	0.272

Table 4.4 shows the results and the graphs are shown in Fig. 4.9.

The stoichiometric air, A_0 (kmol) can be determined from the combustion equation: per kg fuel.

$$\frac{0.83}{12}[C] + \frac{0.16}{2}[H_2] + 0.21A_0[O_2] + 0.79A_0[N_2] \rightarrow$$

$$a[CO_2] + b[N_2] + c[H_2O]$$

therefore, $a = 0.83/12 \qquad c = 0.16/2$
$\qquad\qquad b = 0.79A_0 \qquad a + \tfrac{1}{2}c = 0.21A_0$

Therefore $A_0 = 0.52$ kmol = **15.08 kg/kg**.

The curves show that the maximum *BMEP* occurs with a rich mixture, whereas the minimum *SFC* (giving the best economy) is with a lean mixture.

4.7 Volumetric efficiency

A four cylinder petrol engine of bore 146 mm, stroke 190 mm has a compression ratio of 8:1 and develops 150 kW brake power at 2000 rev/min using a 20 % rich mixture (i.e. air:fuel ratio of stoichiometric/1.2).

The gravimetric analysis of the petrol is 85 % C, 15 % H_2 and the calorific value is 46 MJ/kg. The volumetric efficiency = 70 %, mechanical efficiency = 90 %.

Calculate the indicated thermal efficiency.

Also calculate the air standard efficiency, and outline the reasons why the two efficiencies differ.

Assume the ambient temperature = 15 °C.

Solution Swept volume = $\frac{\pi}{4}(0.146)^2(0.190) \times 4 \times \tfrac{1}{2}(2000/60)$ = 0.212 m³/s. At 15 °C the density of air = $p/RT = \dfrac{1 \times 10^5}{287 \times 288}$ = 1.21 kg/m³, therefore air charge = 0.212 × 1.21 × 0.7 = 0.180 kg/s.

Stoichiometric air:fuel ratio = $\dfrac{100}{21}\left[\dfrac{0.85}{12} + \dfrac{0.15}{2}(\tfrac{1}{2})\right]$ kmol/kg

$= 0.5159 \times 29 = 14.96$ kg/kg.

Therefore, minimum fuel = 0.180/14.96 = 0.012 kg/s

$$\text{ind. thermal efficiency} = \frac{150/0.9}{0.012 \times 1.2 \times 4600} = \mathbf{0.252}$$

$$ASE = 1 - (1/8)^{0.4} = \mathbf{0.564}$$

The differences arise because of the actual properties of the working fluid, and non-adiabatic compression and expansion processes, in the cylinder.

4.8 Supercharging

> A three-litre four-stroke diesel engine develops 780 kJ indicated power per m^3 induced air (at 1 atm, 15 °C). Speed = 3500 rev/min, volumetric efficiency = 80 %. Mechanical efficiency = 80 %.
>
> A mechanically driven supercharger, of isentropic efficiency 70 % and pressure ratio 1.7, is fitted to increase the brake power.
>
> Determine the percentage increase in brake power due to the supercharger.

Solution The swept volume = $\frac{1}{2}(3500)(3 \times 10^{-3})$ = 5.25 m^3/min = 0.0875 m^3/s, therefore naturally aspirated volume = 0.8×0.0875 = 0.07 m^3/s. The supercharger delivery pressure = 1.7 atm, therefore isentropic temperature after compression = $288(1.7)^{0.4/1.4}$ = 335 K, therefore actual temperature = $288 + \left(\dfrac{335 - 288}{0.7}\right)$ = 355 K.

Volume delivered at STP = $\dfrac{1.7}{1.0} \times \dfrac{288}{355} \times 0.0875$ = 0.121 m^3/s, therefore increase in induced volume = $0.121 - 0.070$ = 0.051 m^3/s. Therefore, increase in indicated power = 780×0.051 = 39.8 kW. There is also an increase in the indicated power due to the increased induction pressure, of $(1.7 - 1) \times 1.013 \times 10^5 \times 0.07$ W = 5.0 kW.

Therefore, total increase = 44.8 kW

increase in brake power = $0.8(44.8)$ = 35.8 kW

The power required to drive the supercharger must be deducted to give the net increase.

Supercharger power = $mC_p(355 - 288)$

$$= \frac{1.7 \times 1.013 \times 10^5 \times 0.07}{287 \times 355}(1.005)(355 - 288)$$

$$= 7.9 \text{ kW}$$

Therefore net increase in shaft power = $35.8 - 7.9$ = **27.9 kW**.

Without supercharging, shaft power = $0.8 \times 780 \times 0.07$ W
= 43.7 kW

therefore, % increase = $100(27.9)/43.7$ = **63.8**.

Problems

1 A C.I. engine runs on a constant pressure (Diesel) cycle. Derive an equation for the ideal cycle efficiency in terms of the compression and cut-off ratios.

The condition of the air at the beginning of compression is 0.95 bar, 17 °C. The maximum cycle temperature = 1200 °C, and the heat added at constant pressure = 645 kJ/kg. C_p = 1.15 kJ/kg K.

Calculate the cycle efficiency and *IMEP*.
Answer   64.7 %; 4.41 bar

2   An engine working on the Otto (constant volume) cycle has a heat input = 460 kJ/kg. The air conditions at the beginning of compression are 0.95 bar, 27 °C. The temperature at the end of expansion is 211 °C.

Calculate the compression ratio; maximum pressure and temperature during the cycle; and cycle efficiency. C_p = 1.10 kJ/kg K.
Answer   7.78; 27.08 bar, 1100 K; 56.0 %

3   In an air standard dual combustion cycle the air inlet conditions are 1 bar, 20 °C. Compression ratio = 15:1. Maximum cycle pressure = 70 bar.

The heat added at constant volume = heat added at constant pressure.

Calculate the maximum cycle temperature, *ASE*, and *IMEP*.
Answer   1727 K; 65.4 %; 5.98 bar

4   An I.C. engine runs on an ideal cycle in which the heat rejection is at constant volume, but the heat is supplied such that dp/dv = constant. The heat is supplied over $\frac{1}{8}$ × expansion stroke. Compression ratio = 6:1. Compression commences at 0.95 bar, 20 °C. Maximum cycle pressure = 40 bar.

Calculate the cycle efficiency, *IMEP*, and maximum temperature during the cycle.
Answer   45.1 %; 13.55 bar; 3343 K

5   In a Diesel engine cylinder the temperature of the air is 750 K at the end of compression. At the end of constant pressure combustion the temperature = 2600 K, and for the combustion products C_p = 0.3 + 1 × $10^{-4}T$ kJ/kg, C_v = 0.2 + 1 × $10^{-4}T$ kJ/kg where T = temperature (K). Compression ratio = 15:1. Expansion index = 1.30.

Conditions at the beginning of compression are 1 bar, 17 °C.

Calculate the compression index, and heat transfer during compression; the temperature at the end of expansion and heat rejected during expansion.
Answer   1.35; 47 kJ/kg; 1670 K; 452 kJ/kg

6   A four-stroke Diesel engine runs at 600 rev/min, and develops 150 kW indicated power.

Calorific value of fuel = 44 MJ/kg.

The indicated thermal efficiency = 70 % of the air standard cycle, in which the air conditions at the beginning of compression are 1 bar, 25 °C; the maximum pressure is 65 bar; and the compression ratio is 15.

Estimate a suitable swept volume.

Assume a constant volume air standard cycle. Air:fuel ratio = 30 kg/kg.
Answer 0.0088 m^3

7 In an air standard Otto (constant volume) cycle the compression ratio = 9:1. Compression begins at 0.95 bar, 17 °C. The maximum cycle temperature = 1100 °C. C_v = 0.718 kJ/kg K.

Calculate the heat supplied/kg air, maximum cycle pressure, temperature at the end of expansion, *ASE*, and *IMEP*.

Also calculate the efficiency of a Carnot cycle working between the same temperature limits.
Answer 484.7 kJ; 40.50 bar; 571 K; 58.5 %; 3.63 bar; 78.9 %

8 An engine runs on a dual combustion cycle, with a compression ratio of 20:1. Initial temperature = 13 °C, pressure = 1.0 bar.

Given that the mean index of compression = 1.35 calculate the temperature at the end of compression.

The heat addition is 28 MJ/kmol at constant volume, and 7 MJ/kmol at constant pressure. The mean index of expansion is 1.30.

Calculate the temperature at the end of heat supply, and at the end of expansion; and the thermal efficiency of the cycle.

$$C_v = 18.2 + 3.3 \times 10^{-3} T \text{ kJ/kmol K}$$
$$C_p = 26.3 + 3.9 \times 10^{-3} T \text{ kJ/kmol K}$$

where T = temperature (K).
Answer 816 K; 2240 K, 938 K; 56.7 %

9 A four-stroke aero engine has four cylinders, 140 mm bore, 120 mm stroke. At a speed of 3000 rev/min the volumetric efficiency is 75 %. The air:fuel ratio = 16 kg/kg and the fuel has a calorific value of 44 MJ/kg.

Calculate the mass flow rate of mixture and the indicated power of the engine.

Ambient air conditions = 0.90 bar, 0 °C.
Density of fuel = 760 kg/m^3.
Ind. thermal efficiency = 40 %.

How would you expect the power to change with altitude?
Answer 0.169 kg/s; 175 kW

10 A four-stroke six-cylinder engine has a cylinder of bore 160 mm, stroke 200 mm; clearance volume = 0.6 litres. Calculate the compression ratio and *ASE*.

The engine develops a brake torque of 190 N m at a speed of 3000 rev/min, and the brake thermal efficiency = 26 %.
Fuel:*CV* = 44 MJ/kg.

Calculate the brake power output and *SFC*.
Answer 7.7, 55.7 %; 59.7 kW, 0.314 kg/kWh

11 A four-stroke three-cylinder petrol engine develops 25 kW brake power at a speed of 2400 rev/min.

Analysis of fuel, by mass: 84 % carbon, 15 % hydrogen. The fuel is burned with 20 % excess air.

Calorific value of fuel = 44 MJ/kg. Volumetric efficiency = 80 %. Mechanical efficiency = 80 %. Mean piston speed = 10 m/s. Compression ratio = 9:1.

Assuming that the thermal efficiency = 65 % × air standard efficiency calculate the bore and stroke of the cylinders.
Answer 76 mm, 125 mm

12 A six-cylinder Diesel engine develops 110 kW at 2400 rev/min, with an *SFC* = 0.24 kg/kWh.

Fuel: *CV* = 42 MJ/kg. Analysis, by mass: 86 % C, 14 % H_2. The analysis of the dry exhaust gas, by volume, is 8.9 % CO_2, no CO.

Exhaust gas temperature = 380 °C. Ambient temperature = 20 °C.

The cooling water flow rate = 1 kg/s with a temperature rise of 24 °C.

Draw up an energy balance for the engine.

Assume that for the exhaust gas C_p = 1.15 kJ/kg K.
Answer Brake output 35.7 %, exhaust gas loss 25.0 %, cooling water loss 32.7 %

13 A four-cylinder four-stroke engine develops 45 kW at a speed of 4000 rev/min. Suction pressure = 0.95 bar. Maximum pressure = 60 bar. Compression ratio = 8:1. Brake *SFC* = 0.31 kg/kWh. *CV* of fuel = 44 MJ/kg. The *IMEP* = 60 % of the ideal, constant-volume air standard cycle, *MEP*.

Given that the bore = the stroke, calculate the cylinder bore and mechanical efficiency.
Answer 77.8 %; 103 mm

14 The *SFC* of a four-stroke petrol engine = 0.33 kg/kWh. *CV* of fuel = 44 MJ/kg. Mechanical efficiency = 75 %. Brake power = 30 kW.

Calculate the indicated thermal efficiency. If the air:fuel ratio = 18 kg/kg, volumetric efficiency = 80 %, and the ambient conditions are 1 bar, 20 °C, calculate the *IMEP*. Speed = 2000 rev/min.

The compression ratio = 8:1. Calculate the air standard efficiency and relative efficiency (ind. thermal: *ASE*).
Answer 56.5 %; 58.6 %

15 A C.I. engine operates on a constant-pressure cycle, and is fitted with a supercharger.

Intake is at 1 bar, 15 °C and in the supercharger the pressure ratio = 1.5 with an isentropic efficiency of 60 %.

Air used = 0.25 kg/s. Air:fuel ratio = 25 kg/kg. CV fuel = 43 MJ/kg.

Maximum pressure in the cylinder = 55 bar. The exhaust pressure = 1.6 bar.

The expansion and compression index = 1.30. C_p (air) = 1.005 kJ/kg: (combustion products) = 1.15 kJ/kg K.

Calculate the compression ratio, maximum temperature in the cycle, net indicated power output, and indicated thermal efficiency.
Answer 15.96; 2235 K; 198 kW; 46.0 %

16 A test on a four-cylinder four-stroke oil engine, cylinder bore 250 mm, stroke 300 mm, gave the following results:

duration of test = 30 min
total fuel used = 17.5 kg
CV of fuel = 45 MJ/kg
mean speed = 590 rev/min
mean effective pressure = 6.3 bar
hydraulic brake load = 160 kg
hydraulic brake arm radius = 1.5 m
cooling water flow = 0.50 kg/s
cooling water temperature rise = 48 °C
exhaust gas temperature = 450 °C
air:fuel ratio = 29 kg/kg
C_p (exhaust gas) = 1.15 kJ/kg K
ambient temperature = 15 °C

Draw up an energy balance, showing the percentage of the heat input in brake output, indicated power output, cooling water loss, and exhaust gas loss.
Answer 33.26 %, 41.71 %, 23.00 %, 33.35 %

17 An oil engine running at constant speed gave the results shown in Table 4.5.

Table 4.5

Fuel (kg/s)	exhaust gas analysis, dry, by volume		
	% CO_2	% CO	% O_2
0.065	3.0	0.7	15.5
0.120	6.9	0.4	10.5

Fuel analysis, by mass: 83.3 % C; 13.0 % H_2; 1.3 % O_2.

Calculate the air:fuel ratio and kg/s air used at each load.

Given that the volumetric efficiency at light load is 80 %, estimate its value at the full load.

Answer   53.2 kg/kg, 3.46 kg/s; 27.5 kg/kg, 3.3 kg/s; 76 %

18   A petrol-engine test gave the results shown in Table 4.6.

Table 4.6

Speed (constant) = 600 rev/min

Brake load (N)	125	134	136	137	136	132	121	101
Fuel consumption (cm³/s)	1.30	1.13	1.06	1.00	0.95	0.90	0.83	0.78

For the dynamometer, brake power = $\dfrac{\text{load}(N) \times \text{speed}(\text{rev/min})}{16\,700}$

Fuel: density = 780 kg/m³, CV = 42 MJ/kg.
Air flow measured by orifice: 3.8 cm diameter
$$C_d = 0.6$$
Head (constant) = 0.36 in W.G.
Stoichiometric air:fuel ratio = 14.7 kg/kg
Air density = 1.20 kg/m³

Plot the graphs of brake power, *SFC* and brake thermal efficiency against air:fuel ratio.

Comment on the curves obtained.

19   A 7 MW Diesel engine characteristic is given by

$$I = 4.5 + 1.2L + 0.09L^2 - 0.002L^3$$

where I = heat input (MW), L = brake load (MW).

Plot a graph of *SFC* against load.

The load requirement over one year is as shown in Table 4.7.

Table 4.7

Load (MW)	7	6	5	4	2
Duration (h)	200	500	1000	1000	2300

Given that the fuel cost = 0.20 p/MJ, calculate the annual fuel cost and the mean cost of the generated power (CV fuel = 44 MJ/kg).
Answer   £362 530, 2.01 p/kWh

20   Outline the effects of varying the mixture supplied to a petrol engine on

(a) power output
(b) thermal efficiency

(c) exhaust products composition (including NO_x and unburnt hydrocarbons).

A four-stroke petrol engine has a total swept volume of 1584 c.c. When running at part throttle the fuel consumption is 10.1 l/h, engine speed is 3000 rev/min and the power output is 18 kW. The fuel used can be taken as octane C_8H_{18}, of specific gravity 0.72 and calorific value 44.5 MJ/kg.

Estimate the *BMEP* and brake thermal efficiency.

The mixture strength is weak, there being 10 % excess air. Estimate the volumetric efficiency, based on air consumption only, referred to ambient conditions of 1.01 bar, 288 K.
Answer 6.57 bar, 20.0 %; 69 %

21 A four-stroke, spark ignition engine has six cylinders. Bore = 85 mm, stroke = 75 mm. Compression ratio = 9. Calorific value of fuel = 44 MJ/kg.

A test on the engine gave the following results:

   speed = 5000 rev/min
   output torque = 177.6 Nm
   fuel flow rate = 27.0 kg/h
   air flow = 340 m³/h at ambient conditions
   ambient pressure = 1.10 bar
   ambient temperature = 17 °C
   torque required to motor engine = 18.2 Nm

Calculate the brake and indicated thermal efficiencies, mechanical efficiency, brake *SFC*, air:fuel ratio, and volumetric efficiency.
Answer 0.282, 0.311; 0.907; 0.290 kg/kWh; 16.6 kg/kg; 0.89

5

Gas turbines

Gas turbines are widely used in the conversion of chemical energy into thermal and then mechanical/electrical energy. The main fields of application are in transport (jet engines) and electricity generation, for on-site generation or stand-by purposes. The large amount of heat in the exhaust gases can often be used, by heat-recovery methods, for process requirements. The use for combined heat and power considerably increases the thermal efficiency of the plant.

(Brayton) Joule cycle   The ideal cycle is shown in Fig. 5.1, and the flow diagrams in Fig. 5.2.

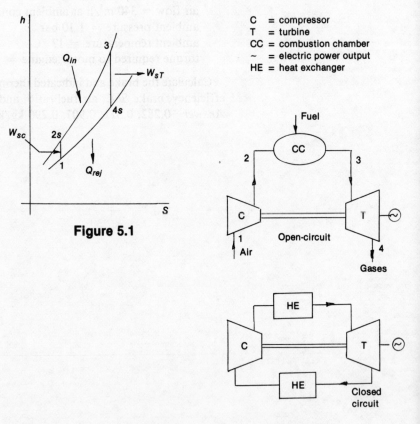

C = compressor
T = turbine
CC = combustion chamber
~ = electric power output
HE = heat exchanger

Figure 5.1

Figure 5.2

72 SOLVING PROBLEMS IN APPLIED THERMODYNAMICS AND ENERGY CONVERSION

For the isentropic compression

$$\frac{T_{2s}}{T_1} = \left(\frac{p_2}{p_1}\right)^{(\gamma-1)/\gamma} = r_p^{(\gamma-1)/\gamma}$$

where r_p = pressure ratio.

The compressor work input $W_{sc} = m_a(h_{2s} - h_1)$.

For the isentropic expansion

$$\frac{T_{4s}}{T_3} = \left(\frac{1}{r_p}\right)^{(\gamma-1)/\gamma}$$

and the turbine work output $W_{st} = m_g(h_3 - h_{4s})$.

The ideal Joule cycle considers an ideal gas and $m_g = m_a$, and the efficiency is then

$$\eta = 1 - \left(\frac{1}{r_p}\right)^{(\gamma-1)/\gamma}$$

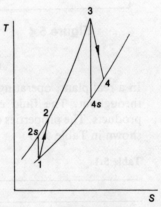

Figure 5.3

Cycle modifications   The irreversibilities in the compressor and turbine flow are allowed for by the use of the isentropic efficiencies. Referring to Fig. 5.3, compressor isentropic efficiency

$$\eta_c = \frac{h_{2s} - h_1}{h_2 - h_1}$$

turbine isentropic efficiency

$$\eta_t = \frac{h_3 - h_4}{h_3 - h_{4s}}$$

cycle efficiency $= 1 - \dfrac{h_4 - h_1}{h_3 - h_2}$

net work output $= (h_3 - h_4) - (h_2 - h_1)$.

The compressor work can be reduced by the use of multi-stage compression, with intercooling between stages, as shown in Fig. 5.4.

The turbine work, as in the case of steam turbines, can be considerably increased by the use of reheat. The gases are reheated between stages, as shown in Fig. 5.5.

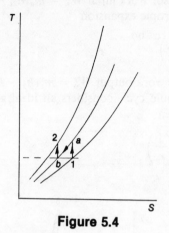

Figure 5.4

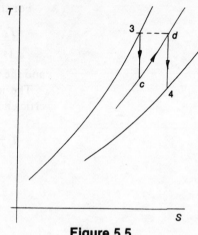

Figure 5.5

Practical 'cycles'

In a real plant, operating on a flow circuit, the working fluid is not air throughout. The fluid expanding through the turbine is combustion products. The properties can be allowed for by the use of the mean values shown in Table 5.1.

Table 5.1

	C_p (kJ/kg K)	$\gamma = C_p/C_v$
Air	1.005	1.40
Combustion products	1.15	1.33

In addition, the specific heat of a real gas varies with temperature; there are pressure drops between the various parts of the plant and in the combustion chamber, and the mass flow rate is not constant. The mass flow rate of combustion products $m_g = m_a$ (air flow) $+ m_f$ (fuel flow). However the air:fuel ratio is normally high (40 to 70 kg/kg) so that $m_g \doteqdot m_a$.

5.1 Joule cycle

Show that in an ideal Joule cycle the net work output is a maximum if $r_p = (\eta_c \eta_T K)^n$ where r_p = pressure ratio, K = maximum temperature/minimum temperature, and $n = \gamma/[2(\gamma - 1)]$.

Determine the efficiency at this condition in terms of β, η_c and K, where $\beta = r_p^{(\gamma-1)/\gamma}$.

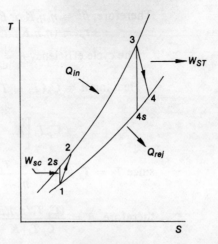

Figure 5.6

Solution Referring to Fig. 5.6 in the ideal (air standard) cycle the working fluid is an ideal gas.

Compressor work $W_{sc} = C_p(T_2 - T_1)$

$$= C_p(T_{2s} - T_1)/\eta_c$$

$$= \frac{C_p T_1}{\eta_c}(r_p^{(\gamma-1)/\gamma} - 1)$$

where $r_p = p_2/p_1$.

Turbine work $W_{st} = C_p(T_3 - T_4)$

$$= C_p \eta_t (T_3 - T_{4s})$$

$$= C_p \eta_t T_3 \left[1 - \left(\frac{1}{r_p}\right)^{(\gamma-1)/\gamma}\right]$$

Therefore net work $W_s = C_p \eta_t T_3 \left[1 - \left(\frac{1}{r_p}\right)^{(\gamma-1)/\gamma}\right]$

$$- \frac{C_p T_1}{\eta_c}(r_p^{(\gamma-1)/\gamma} - 1)$$

$$= \frac{C_p T_1}{\eta_c}\left[\eta_c \eta_t K \left(1 - \frac{1}{\beta}\right) - (\beta - 1)\right]$$

$$= \frac{C_p T_1}{\eta_c}(\beta - 1)\left(\eta_t \frac{\eta_c}{\beta} K - 1\right)$$

where $\beta = r_p^{(\gamma-1)/\gamma}$.

For maximum work $d_{Ws}/d\beta = 0$, therefore

$$(\beta - 1)\left(-\frac{\eta_t \eta_c K}{\beta^2}\right) + \left(\frac{\eta_t \eta_c K}{\beta^2} - 1\right) = 0$$

GAS TURBINES 75

Therefore, $\beta^2 = \eta_c \eta_t K$ or $\beta = (\eta_c \eta_t K)^{\frac{1}{2}}$, therefore
$r_p = (\eta_c \eta_t K)^{\gamma/2(\gamma-1)}$

The cycle efficiency η = net work/heat input

Heat input $= C_p(T_3 - T_2) = C_p T_1 \left(\dfrac{T_3}{T_1} - \dfrac{T_2}{T_1} \right)$

$= C_p T_1 \left[K - 1 - \left(\dfrac{\beta - 1}{\eta_c} \right) \right]$

since $T_2 = T_1 + \left(\dfrac{T_{2s} - T_1}{\eta_c} \right) = T_1 + \dfrac{(\beta - 1)}{\eta_c} T_1$

therefore, $\eta = \dfrac{(C_p T_1/\eta_c)(\beta - 1)(\eta_t \eta_c K/\beta - 1)}{C_p T_1 [K - 1 - (\beta - 1)/\eta_c]}$

$= \dfrac{(\beta - 1)(\eta_t \eta_c K - \beta)}{\eta_c \beta [(K - 1) - (\beta - 1/\eta_c)]}$

Thus at the condition of maximum output $\beta = (\eta_c \eta_t K)^{\frac{1}{2}}$ substitution gives

$\eta = \dfrac{(\beta - 1)(\beta^2 - \beta)}{\eta_c \beta [(K - 1) - (\beta - 1)/\eta_c]} = \dfrac{(\beta - 1)^2}{\eta_c(K - 1) - (\beta - 1)}$

5.2 Open cycle

> An open-cycle gas turbine plant consists of a compressor, combustion chamber and turbine.
> Assuming that the working fluid is a perfect gas, show that for a net work output $\eta_c \eta_t T_3 > T_1 r_p^{(\gamma-1)/\gamma}$, where T_1, T_3 are the minimum and maximum temperatures, and r_p = pressure ratio.
> The inlet conditions to the compressor are 1 bar, 15 °C and the compressor isentropic efficiency η_c = 80 %. The inlet conditions at the turbine are 4 bar, 620 °C, and the turbine isentropic efficiency = 85 %.
> Calculate the net work, work ratio and cycle efficiency.

Solution Referring to Fig. 5.6, compressor work $W_{sc} = C_p(T_2 - T_1)$

$= \dfrac{C_p}{\eta_c}(T_{2s} - T_1)$

$= \dfrac{C_p T_1}{\eta_c}(r_p^m - 1)$

where $m = \dfrac{\gamma - 1}{\gamma}$.

The turbine work $W_{st} = C_p(T_3 - T_4) = \eta_t C_p(T_3 - T_{4s})$
$= \eta_t C_p T_3(1 - 1/r_p^m)$

The net work = $\eta_t C_p T_3(1 - 1/r_p^m) - \dfrac{C_p T_1}{\eta_c}(r_p^m - 1)$ and will be positive if

$$\eta_t T_3(1 - 1/r_p^m) > \dfrac{T_1}{\eta_c}(r_p^m - 1)$$

i.e. $\eta_c \eta_t T_3 (r_p^m - 1)/R > T_1(r_p^m - 1)$
i.e. $\eta_c \eta_t T_3 > T_1 r_p^m$

$T_1 = 288$ K, therefore $T_{2s} = 288(p_2/p_1)^m = 288(4)^{0.2857} = 428$ K
therefore $T_2 = 288 + (428 - 288)/\eta_c = 463$ K
$T_3 = 893$ K, therefore $T_{4s} = T_3/4^{0.2857} = 601$ K
therefore $T_4 = 893 - \eta_t(893 - 601) = 645$ K

Compressor work $W_{sc} = 1.005(T_2 - T_1) = 175.9$ kJ/kg and the turbine work $W_{st} = 1.005(T_3 - T_4) = 249.2$ kJ/kg.
Net work = 249.2 - 175.9 = **73.3 kJ/kg**
Heat input = $1.005(T_3 - T_2) = 432.2$ kJ/kg
therefore, cycle efficiency = 73.3/432.2 = **0.170**
Work ratio = net work/turbine work = 73.3/249.2 = **0.294**.

5.3 Open cycle

> An open-cycle gas turbine has an overall pressure ratio of 8:1 and the maximum temperature = 900 K. The ambient conditions are 1 bar, 20 °C. The isentropic efficiencies are 80 % for the compressor, 85 % for the turbine.
> Air flow rate $\doteq$ 15 kg/s. Calorific value of fuel = 42 MJ/kg. Estimate the net work output, cycle efficiency and air:fuel ratio.
> Comment on the values obtained.

Solution Referring to Fig. 5.3, the temperature at the various points in the cycle are calculated.
$T_{2s} = T_1 r_p^{(\gamma-1)/\gamma} = 293(8)^{0.2857} = 531$ K, therefore $T_2 = 293 + (531 - 293)/0.8 = 591$ K. $T_{4s} = T_3/r_p^{(\gamma-1)/\gamma} = 900/8^{0.2857} = 535$ K, therefore $T_4 = 900 - 0.85(900 - 535) = 590$ K.

Compressor work
$$W_{sc} = m_a C_p (T_2 - T_1)$$
$$= 15(1.005)(591 - 293) = 4429 \text{ kW}.$$

Turbine work
$$W_{st} = m_g(1.15)(900 - 590)$$
$$= 356.5 m_g \text{ kW}$$

where m_g = gas flow rate = $m_a + m_f = m_f(A/F + 1)$, and A/F = air:fuel ratio (kg/kg).
Now the

heat input = $m_g h_3 - m_a h_2$
$= m_g(1.15)(900) - 15(1.005)(591) = 1035 m_g - 8909$ kW.

Therefore assuming a combustion efficiency of 100 %, the calorific value of the fuel is converted into this heat input, therefore $m_f = (1035 m_g - 8909)/42\,000$ kg/s $= 0.0246 m_g - 0.2121$.

But $m_g = m_a + m_f = 15 + m_f$.

Therefore substituting $m_f = 0.369 + 0.0246 m_f = 0.2121$, therefore $m_f = 0.161$ kg/s.

Hence turbine work $= 356.5 m_g = 356.5(15.161) = 5405$ kW,

net work $= 5405 - 4492 =$ **913 kW**

cycle efficiency $= 913/(1035 m_g - 8909) = 913/6783 =$ **0.134**

air:fuel ratio $= 15/0.161 =$ **93.2**.

The maximum temperature of 900 K is rather lower than those that can be used in modern plant. Long-life industrial plant can now utilize maximum temperatures of 1200 K, and a little higher in aircraft engines with the use of blade cooling. The raising of the maximum temperature will considerably increase the cycle efficiency.

The modest inlet temperature is mirrored in the high air:fuel ratio, greater than the usual range of values. The efficiency of the plant could be increased by utilization of the waste heat in the gases leaving the turbine in this problem amounting to

$$m_g C_p (T_4 - T_1) = 15.161(1.15)(590 - 293)$$
$$= 5178 \text{ kW}.$$

This represents

$5178 \times 100/6783 = 76.3$ % of the heat input – a substantial loss.

5.4 Reheat cycle with heat exchanger

In a gas-turbine plant the air is compressed from 1 bar, 15 °C to a pressure of 6 bar (isentropic efficiency = 0.80). The combustion products enter the turbine at 6 bar, 1000 K and expands in two stages, each of expansion ratio $\sqrt{6}$, with the gases reheated to 1000 K between stages.

Calculate the net work output/kg air, work ratio and cycle efficiency.

Determine these values if a heat exchanger of effectiveness 0.75 is fitted to the plant: and compare the gas temperature leaving the plant in each case.

Solution Referring to Fig. 5.7,

$T_{2s} = T_1(6)^{0.4/1.4} = 481$ K

$T_2 = 288 + \dfrac{481 - 288}{\eta_c} = 529$ K

$T_{4s} = T_3/6^{0.33/1.33} = 799$ K

$T_4 = 1000 - 0.85(1000 - 799) = 829$ K

$T_5 = T_3 = 1000$ K

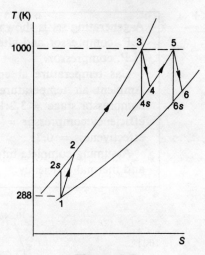

Figure 5.7

$T_{6s} = T_5/6^{0.33/1.33} = 799$ K
$T_6 = \mathbf{829}$ **K**

Compressor work $W_{sc} = 1.005(T_2 - T_1) = 242.2$ kJ/kg
Turbine work $W_{st} = 1.15(T_3 - T_4) + 1.15(T_5 - T_6) = 393.3$ kJ/kg
Heat input $Q_{in} = 1.15 T_3 - 1.005 T_2 = 618.3$ kJ/kg
Net work $= W_{st} - W_{sc} = \mathbf{151.1}$ **kJ/kg**
Work ratio $= 151.1/393.3 = \mathbf{0.384}$
Cycle efficiency $= 151.1/618.3 = \mathbf{0.244}$

The incorporation of a heat exchanger enables some of the heat in the exhaust gases to be transferred to the compressed air before entering the combustion chamber. If the air temperature is raised to T_2' the heat input is reduced to $1.15 T_3 - 1.005 T_2'$.

The amount of heat transferred is measured by the heat exchanger effectiveness, e. Neglecting the difference between the air and gas flow rates,

$$e = \frac{T_2' - T_2}{T_6 - T_2}; \quad \text{therefore } T_2 + e(T_6 - T_2) = T_2'$$

Therefore, $T_2' = 529 + 0.75(829 - 529) = 754$ K.

The net work output and work ratio remain the same, but the cycle efficiency is now

$$151.1/392.2 = \mathbf{0.385}$$

The exhaust gas temperature without a heat exchanger is $T_6 = 829$ K. With a heat exchanger, an energy balance gives

heat absorbed by air = heat from gases
$1.005(754 - 529) = 1.15(829 - T_6)$
$T_6 = \mathbf{632}$ **K**.

5.5 Intercooling + reheat + heat-exchanger cycle

A generating set is shown in Fig. 5.8. The L.P. turbine drives the L.P. compressor and alternator, and the H.P. turbine drives the H.P. compressor.

Gas temperature at entry to both turbine stages = 600 °C. Ambient air temperature = 20 °C. The pressure ratio in each compressor stage = 3.2:1. Mass flow rate = 115 kg/s. Isentropic efficiency:compressor = 80 %, turbine = 85 %. Heat exchanger effectiveness = 0.7.

Assuming complete intercooling calculate the net power output and thermal efficiency.

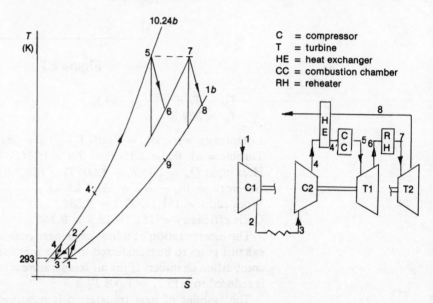

Figure 5.8

Solution The temperature at each point in the cycle is calculated. $T_{2s} = 293(3.2)^{0.4/1.4} = 408$ K, $T_2 = 293 + (408 - 293)/0.8 = 437$ K.

For complete intercooling $T_3 = T_1 = 293$ K.

Also the pressure ratio $p_4/p_3 = p_2/p_1 = 3.2$, therefore $T_4 = T_2 = 437$ K. Compressor work = $115(1.005)(437 - 293) = 16\,640$ kW/stage. Now the H.P. stage C2 is driven by the H.P. turbine T2, therefore the work output from the turbine = 16 640 kW.

It is this requirement that determines the H.P. stage exit pressure p_6.

$16\,640 = 115(1.15)(T_5 - T_6)$, therefore $T_5 - T_6 = 126$, therefore $T_6 = 747$ K. Hence $T_5 - T_{6s} = T_5 - T_6/\eta_t = 126/0.85 = 148$ K giving $T_{6s} = 725$ K. The isentropic relationship can now be used:

$$\frac{T_5}{T_{6s}} = \left(\frac{p_5}{p_6}\right)^{0.33/1.33} = \frac{873}{725}, \text{ therefore } p_5/p_6 = 2.10$$

therefore $p_6 = 4.88$.

Considering the L.P. turbine T1:

$$\frac{T_7}{T_{8s}} = \left(\frac{p_7}{p_8}\right)^{\gamma-1/\gamma} = 4.88^{0.33/1.33} = 1.486$$

Therefore $T_{8s} = 587$ K, therefore

$$T_8 = 873 - 0.85(873 - 587) = 631 \text{ K}.$$

Hence the turbine work $W_{st} = 115(1.15)(873 - 631) = 32\,005$ kW. The L.P. compressor work = 16 640 kW.

Therefore, net power = **15 365 kW**.

The heat exchanger raises the air temperature, and the effectiveness =

$$0.7 = \frac{T'_4 - T_4}{T_8 - T_4}, \text{ therefore}$$

$$T'_4 = 437 + 0.7(631 - 437) = 573 \text{ K}.$$

Hence

$$\text{heat input} = 115(1.15\,T_5 - 1.005\,T'_4) + 115(1.15)(T_7 - T_6)$$
$$= 63\,850 \text{ kW}.$$

Therefore, cycle efficiency = $15\,365/63\,850 = $ **0.24**.

5.6 Aero engine cycle

In a bypass aero engine air at 300 K is compressed in a fan through a pressure ratio of $1\frac{1}{2}$:1, with an isentropic efficiency of 85 %. Part of the flow is then bypassed to the turbine exhaust and the remainder is compressed in the H.P. compressor, through a pressure ratio of 8:1, with an isentropic efficiency of 80 %. The gas temperature at the turbine inlet is 1500 K. The isentropic efficiency of the turbine is 90 %.

Assuming that the turbine exhaust and bypass pressures are equal, and the flow rate through the H.P. compressor and turbine are equal, calculate the proportion of the inlet flow that is bypassed.

Given that the mixing is adiabatic estimate the temperature of the exhaust stream.

Solution The flow and T–s diagrams are shown in Fig. 5.9.

For convenience take $p_1 = 1$ bar, therefore $p_2 = 1\frac{1}{2}$ bar, $p_3 = 1\frac{1}{2} \times 8 = 12$ bar. Also $p_5 =$ bypass pressure $p_2 = 1\frac{1}{2}$ bar.

$$T_{2s} = T_1(1.5)^{0.4/1.4} = 337 \text{ K, therefore}$$
$$T_2 = T_1 + (T_{2s} - T_1)/\eta_c = 344 \text{ K}$$
$$T_{3s} = T_2(8)^{0.2857} = 623 \text{ K, therefore}$$
$$T_3 = T_2 + (T_{3s} - T_2)/\eta_c = 693 \text{ K}$$
$$T_4 = T_{5s}(12/1.5)^{0.33/1.33} = 1500 \text{ K, therefore}$$
$$T_{5s} = 892 \text{ K, and } T_5 = T_4 - 0.9(T_4 - T_{5s}) = 953 \text{ K}.$$

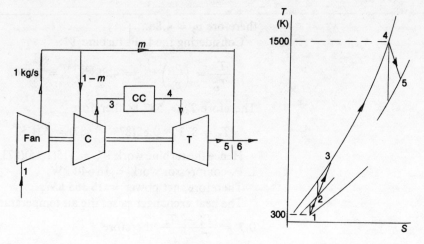

Figure 5.9

The work output from the turbine is used, in this case of an aero engine, to drive the compressor and fan.

Hence,

$$(1 - m)(h_4 - h_5) = (1 - m)(h_3 - h_2) + m(h_2 - h_1),$$

therefore,

$$(1 - m)(1.15)(1500 - 953) = (1 - m)(1.005)(693 - 344) + 1.005 m(344 - 300),$$

giving $m = 0.86$.

Therefore **86 %** of the flow is bypassed.

Considering the exhaust stream $mh_2 + (1 - m)h_5 = h_6$ since the flow at point 6 = 1 kg/s, therefore

$$0.86(1.005)(344) + 0.14(1.15)(953) = 1.15(T_6)$$

giving $T_6 = \mathbf{392\ K}$.

5.7 Variation of specific heat with temperature

In a turbo jet engine a compressor of pressure ratio 10:1 and isentropic efficiency 85 % is driven by a turbine of isentropic efficiency 90 %. The temperature at entry to the compressor and turbine are 300 K and 1600 K respectively. The specific heat is given by: air $C_p = 0.00015\,T + 0.96$ kJ/kg K; combustion products $C_p = 0.00014\,T + 1.06$ kJ/kg K, where T = temperature (K).

Assuming that the isentropic law pv^γ = constant can be used, with mean values of $\gamma = 1.39$ for air and $\gamma = 1.26$ for combustion products, determine the ratio of the turbine exit pressure to the compressor inlet pressure.

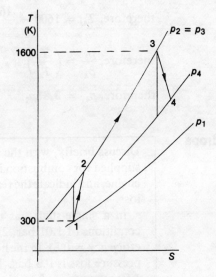

Figure 5.10

Solution Referring to Fig. 5.10, $T_{2s} = 300(10)^{0.39/1.39} = 572$ K, therefore
$$T_2 = 300(572 - 300)/0.85 = 620 \text{ K}.$$

The compressor work $= \int_{T_2}^{T_2} C_p \, dT$

and in this problem, the specific heat is a function of the temperature. The compressor work is

$$W_{sc} = \int_{300}^{620} (0.00015\,T + 0.96) \, dT$$
$$= [0.000075\,T^2 + 0.96\,T]_{300}^{620}$$
$$= 22.08 + 307.2$$
$$= 329.3 \text{ kJ/kg}$$

For a turbo jet engine, the turbine work = compressor work, therefore
$329.3 = \int C_p \, dT$

$$= \int_{T_4}^{1600} (0.00014\,T + 1.06) \, dT$$
$$= [0.00007\,T^2 + 1.06\,T]_{T_4}^{1600}$$
$$= 179.2 + 1696.0 - 0.00007\,T_4^2 - 1.06\,T_4$$

therefore $0.00007\,T_4^2 + 1.06\,T_4 - 1545.9 = 0$
so $T_4 = 1340$ K.

Having determined T_4, the isentropic temperature T_{4s} can be calculated, and hence the pressure p_4.

$$\eta_t = 0.90 = \frac{T_3 - T_4}{T_3 - T_{4s}}$$

GAS TURBINES 83

therefore, $T_{4s} = 1600 - \dfrac{1600 - 1340}{0.9} = 1311$ K

therefore, $\dfrac{p_3}{p_4} = \left(\dfrac{T_3}{T_{4s}}\right)^{\gamma/(\gamma-1)} = (1600/1311)^{1.26/0.26} = 2.626$

therefore, $p_4 = \mathbf{3.81 p_1}$

5.8 Pressure drops

> Discuss briefly, with the aid of a sketch, the distribution of the air supplied for combustion in a can-type combustion chamber for a jet engine, and indicate the reasons for the combustion-chamber pressure loss.
>
> In a single-shaft gas turbine, air is compressed from ambient conditions of 1.01 bar, 25 °C, to a pressure of 10.1 bar (isentropic efficiency = 0.8). In the heat exchanger and combustion chamber the pressure loss is 0.3 bar. The turbine entry temperature = 850 °C. After expansion through the turbine the gases pass through the heat exchanger (effectiveness, or thermal ratio = 0.76), and the gas side pressure loss = 0.15 bar. The gases then pass to atmosphere. Turbine isentropic efficiency = 0.84.
>
> Neglecting the effect of fuel addition on mass flow through the system, and any pressure losses other than those mentioned, estimate the air mass flow rate and cycle efficiency for a net power output of 15 MW.

Solution The combustion chamber is required to burn large amounts of fuel with high air:fuel ratios (of the order 45 to 120 kg/kg). The heat release has to be a maximum in the limited volume available, and a smooth stream of uniformly heated gases is needed for expansion through the turbine, with the minimum pressure loss.

Air from the compressor enters the combustion chamber at a velocity up to 150 m/s and must be decelerated to avoid blow-off of the flame. The flame is lit in a region of low axial velocity, so that it remains alight at all engine operating conditions.

A typical air flow distribution is shown in Fig. 5.11.

In the burning zone, swirl vanes produce a strong toroidal vortex, and combustion products are recirculated to reduce the combustion time of fresh fuel. The conical fuel spray from the burner intersects the recirculation vortex at its centre: this, together with the turbulence in the primary zone, considerably increases the mixing of the fuel and air.

A pressure loss is incurred in providing the high turbulence and adequate mixing.

$$T_{2s} = T_1(10)^{0.4/1.4} = 575 \text{ K}$$
$$T_2 = T_1 + (T_{2s} - T_1)/\eta_c = 645 \text{ K}$$
$$T_{4s} = T_3 / \left(\dfrac{9.8}{1.16}\right)^{0.33/1.33} = 659 \text{ K}$$

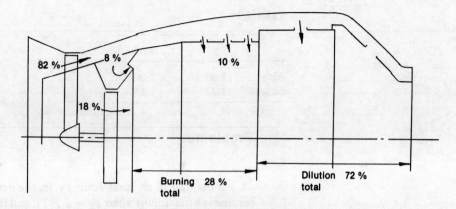

Figure 5.11

$$T_4 = T_3 - \eta_t(T_3 - T_{4s}) = 733 \text{ K}$$
$$e = (T_5 - T_2)/(T_4 - T_2) = 0.76$$

therefore, $T_5 = 629$ K.

Net work $= 1.15(T_3 - T_4) - 1.005(T_2 - T_1)$
$= 448.5 - 348.7$
$= 99.8$ kJ/kg

Therefore, air flow rate $= 15\,000/99.8 = \mathbf{150.3}$ **kg/s**

$$\text{Cycle efficiency} = \frac{\text{net work}}{h_3 - h_5} = \frac{99.8}{1.15(1123) - 1.005(629)} = \mathbf{0.15}.$$

Unless stated otherwise, the values shown in Table 5.2 are used in the exercises.

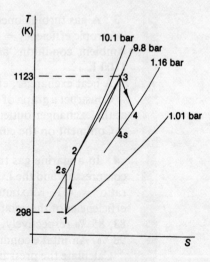

Figure 5.12

Table 5.2

	γ	C_p (kJ/kg K)
Air	1.40	1.005
Gases	1.33	1.15

Problems

1 Show that in an ideal Joule cycle, the net output from the gas turbine is a maximum when $T_2 = \sqrt{T_1 T_3}$, and that the cycle efficiency depends only upon the ratio $T_2:T_1$.
T_2 = temperature at end of compression
T_1 = inlet temperature to compressor
T_3 = inlet temperature to turbine
Given that a heat exchanger of effectiveness e is used in the cycle derive an expression for the cycle efficiency in terms of e, the pressure ratio $\alpha = \left(\dfrac{p_2}{p_1}\right)^{(\gamma-1)/\gamma}$, and temperature ratio $K = T_3/T_1$.

2 A gas turbine operates on a pressure ratio of 8:1. The ambient conditions are 1 bar, 20 °C. Maximum temperature = 1100 K.
Calculate the net work and cycle efficiency for the following cases:

(a) compressor and turbine isentropic efficiencies of 100 %;
(b) compressor isentropic efficiency = 90 %, turbine isentropic efficiency = 85 %.

Answer (a) 274 kJ/kg, 37.4 %; (b) 180 kJ/kg, 16.7 %

3 A gas turbine operates on a pressure ratio of 8:1. Compressor isentropic efficiency = 80 %. Turbine isentropic efficiency = 85 %. Ambient conditions are 1 bar, 20 °C. Maximum temperature = 1100 K.
A heat exchanger, effectiveness e, is fitted into the system.
Construct a graph of net work, cycle efficiency and gas temperature at the exchanger outlet plotted against the effectiveness.
Comment on the curves.

4 In a marine gas turbine unit the H.P. turbine stage drives the compressor, and the L.P. stage drives the propellor. Overall pressure ratio = 4:1. Maximum temperature = 650 °C. The isentropic efficiencies of the compressor, H.P. turbine and L.P. turbine are 80, 83, 85 % respectively. The mechanical efficiency of both shafts is 98 %. Air intake conditions 1.01 bar, 25 °C. Mass flow = 60 kg/s.
Calculate the pressure at the H.P. turbine exit, thermal efficiency and shaft power.

Answer 1.60 bar, 14.1 %, 4800 kW

5 A gas turbine plant uses reheat between the turbine stages.

Show that the ratio of the reheat fuel flow to the turbine exhaust gas flow, α_R, is given by

$$\alpha_R = \Delta h_g / ECV$$

where Δh_g = enthalpy increase/kg of the turbine gases in the reheat process.

$$ECV = \text{net } CV + \frac{h_{a2}}{\alpha_s} - \left(1 + \frac{1}{\alpha_s}\right) h_{s2}$$

α_s = stoichiometric fuel:air ratio
h_{a2} = enthalpy of air/kg at the reheat exit temperature
h_{s2} = enthalpy of the stoichiometric products/kg at the reheat exit temperature.

Assume that there are no combustion losses, and that the enthalpy of the fuel is negligible.

Also show that the total fuel:air ratio, α_2, at the reheat system exit is given by:

$$\alpha_2 = \alpha_s + \alpha_R(1 + \alpha_s)$$

6 A *closed* cycle consists of a compressor, heater, turbine and cooler. The air temperature = 30 °C at the compressor inlet, 800 °C at the turbine inlet. Mass flow rate = 20 kg/s. Isentropic efficiencies: compressor = 80 %, turbine = 85 %.
$C_p = 1.00$ kJ/kg K. $\gamma = 1.40$.

Determine the pressure ratio for maximum power output, and at these conditions the net power output and thermal efficiency.
Answer 4.65; 2.31 MW, 20.6 %

7 In an open circuit gas turbine set the pressure ratio = 5:1. Isentropic efficiency: compressor 80 %, turbine 85 %. Compressor inlet conditions 1.0 bar, 15 °C. Air flow rate = 15 kg/s, fuel flow rate = $\frac{1}{4}$ kg/s. CV of fuel = 42 MJ/kg.

Determine the gas temperature at the turbine inlet, net work output, and thermal efficiency.

The set is fitted with a heat exchanger so that the turbine inlet temperature is increased to 1120 K. If the air and fuel flow rates remain the same determine the heat exchanger effectiveness.
Answer 1027 K, 1.9 MW, 18.1 %, 0.36

8 An open circuit gas turbine plant comprises:

(a) A two-stage compressor, pressure ratio 16:1, with complete intercooling between stages. Isentropic efficiency of each compressor stage = 85 %.

(b) A two-stage turbine with reheat between stages to 700 °C. The gas temperature at the H.P. inlet = 800 °C. Isentropic efficiency of each turbine stage = 90 %.

(c) A heat exchanger of effectiveness = 0.75.

Inlet conditions at the compressor are 1 bar, 20 °C.

Calculate the gas pressure at the H.P. turbine exit, specific work output, cycle efficiency and gas temperature leaving the exchanger.

The H.P. turbine drives both compressor stages.

Answer 3.74 bar, 283 kJ/kg, 36.7 %, 550 K

9 A closed cycle gas turbine uses helium ($\gamma = \frac{5}{3}$) as the working fluid. Isentropic efficiencies: compressor = 85 %, turbine 88 %. Maximum temperature = 870 K. The temperature at the compressor inlet = 15 °C.

Determine the pressure ratio at which the specific work output is a maximum.

Would this pressure ratio give the maximum efficiency? If not, at what pressure ratio would the efficiency be a maximum?

Answer 2.76, No, 1.46

10 The maximum temperature in an aircraft jet engine is 800 °C, and the pressure ratio = 5:1. The isentropic efficiencies are: compressor 86 %, turbine 89 %. The ambient conditions are 1 bar, 17 °C. The air consumption when the aircraft is stationary is 13.6 kg/s.

CV of fuel = 44 MJ/kg.

The gases are discharged through an area of 0.01 m². Calculate the air:fuel ratio, gas exhaust velocity and thrust developed.

Answer 3.6 kg/kg; 625 m/s; 9780 N

11 Helium is used as the working fluid in an ideal Joule cycle. Gas enters the compressor at 20 bar, 27 °C and leaves at 60 bar. The gas is then heated to 850 °C in a heat exchanger before expansion through a turbine and passage through a precooler.

Determine:

(a) the temperature of the helium at the end of the expansion and compression processes,

(b) the heat supplied, heat rejected and net work/kg of helium,

(c) the thermal efficiency of the cycle and the work ratio.

For helium C_p = 5.23 kJ/kg K, $\gamma = C_p/C_v$ = 1.64.

Answer 461 K, 732 K; 3462, 2224, 1238 kJ/kg; 0.36, 0.60

12 Air enters an industrial gas turbine at 1 bar, 300 K and is compressed to 6 bar in an L.P. compressor unit. The air leaving is divided into two streams A and B.

The stream A is further compressed to 24 bar in an H.P. compressor, heated in a combustion chamber to 1300 K, and then expands in a turbine to 1 bar before exhausting to the atmosphere. This turbine is coupled by a shaft to both the L.P. and H.P. compressors.

The stream B is preheated in a heat exchanger, and then finally heated to 1300 K in a combustion chamber. The hot gases then expand in a free power turbine to 1 bar and exhaust to the atmosphere after passing through the gas side of the heat exchanger. Thermal ratio of the heat exchanger = 0.7.

Isentropic efficiencies: both compressors 0.84
both turbines 0.86

Determine the ratio of the mass flow rates and the thermal efficiency of the engine.
Answer $A:B$ = 1.59; 0.27

13 For a simple gas turbine cycle it can be shown that the pressure ratio r_p must lie between 1 and $(T_3/T_1)^{\gamma/(\gamma-1)}$, and the specific work output is a maximum when $r_p = (T_3/T_1)^{\gamma/2(\gamma-1)}$, where T_1 and T_3 are the compressor and turbine inlet temperatures.

With the aid of a T–s diagram explain the physical significance of these conclusions.

Show that if, in practice, the isentropic efficiencies of the component and turbine are η_c, η_t respectively, maximum specific work will occur when $r_p = (\eta_c \eta_t T_3 / T_1)^{\gamma/2(\gamma-1)}$.

Hence calculate the maximum specific work output for the plant when T_1 = 300 K, T_3 = 850 K, η_c = 0.84, η_t = 0.90 and C_p = 1.005 kJ/kg K throughout. γ = 1.40.

How would you describe the cycle if it operates at a pressure ratio $r_p = (\eta_c \eta_t T_3/T_1)^{\gamma/(\gamma-1)}$?
Answer 77.1 kJ/kg

14 An open-cycle gas turbine unit is to be used as a stand-by generator with an output of 250 kW. The inlet conditions to the compressor are 1 bar, 27 °C and the compressor pressure ratio = 7, isentropic efficiency = 0.8. Combustion chamber pressure loss = 0.4 bar. Combustion efficiency = 0.96. Inlet temperature to H.P. turbine = 850 °C. Isentropic efficiency of H.P. turbine = 0.88, L.P. turbine = 0.86. The pressure loss in the L.P. turbine exhaust duct = 0.2 bar.

Given that the calorific value of the fuel is 40 MJ/kg, determine the L.P. turbine pressure ratio, air and fuel mass flow rates, and overall thermal efficiency.
Answer 2.80; 1.28, 0.02 kg/s; 0.35

15 A simple turbo jet propulsion unit was tested on a static bed with ambient conditions of 0.99 bar, 15 °C.
Compressor: stagnation pressure ratio = 4.2:1
stagnation isentropic efficiency = 0.83

Turbine: stagnation isentropic efficiency = 0.92
Combustion efficiency = 0.98
Air flow rate = 18 kg/s
Combustion chamber pressure loss = 0.06 bar
Maximum cycle stagnation temperature = 1050 K
Calorific value of fuel = 40 MJ/kg
Air:fuel ratio = 50 kg/kg

The propulsion nozzle is designed to produce critical flow (sonic conditions) at the exit, and the efficiency = 0.92.

The critical temperature ratio = $(\gamma + 1)/2$.

Determine the air:fuel ratio, nozzle exit area, design thrust, and specific fuel consumption, kg/h per N thrust.

Answer 51.4 kg/kg, 0.066 m², 10 050 N, 0.122

16 Outline the effects of using (a) reheat, (b) a heat exchanger, in a gas turbine plant, for aero engine and electricity generation applications.

A method of increasing the mean temperature of heat addition to a compressor–combustion-chamber–turbine–heat-exchanger cycle is to bleed off a fraction $\Delta m/m$ of the air delivered by the compressor and using it to cool the turbine blades. The maximum cycle temperature is thereby increased from T to $T + \Delta T$.

Assuming no pressure loss in the combustion chamber or heat exchanger, the working fluid throughout is air, the bled air does no work in the turbine, and the air temperature entering the combustion chamber is equal to that at the turbine exhaust, show that there is no net gain in efficiency when

$$\frac{\Delta m}{m} = \frac{\Delta T/T}{1 + \Delta T/T}$$

6

Turbomachinery: reciprocating machinery

Machinery is widely encountered in many fields of energy conversion where thermal/potential/kinetic energy is converted into mechanical/electrical power, or vice versa. The use of machinery dates back to ancient times, though limited by the lack of technological knowledge.

Classification of machinery

There are many different kinds of machine (rotating and reciprocating) in use today, and a classification is detailed as follows.

Incompressible-flow machines:

(a) *rotodynamic*
- turbines (radial, mixed flow, axial)
- hydraulic couplings
- pumps (centrifugal, mixed flow, axial)
- propellers, windmills

(b) *positive displacement*
- pumps (multiple axial, radial cylinder, reciprocating, rotary)
- motors (gear, piston, vane)

(c) *miscellaneous*
- hydraulic ram
- ejectors

Compressible-flow machines:

(a) *rotodynamic*
- turbines (impulse, reaction, radial-flow, axial)
- compressors (centrifugal, axial)

(b) *reciprocating*.

Machine parameters: specific speed

For incompressible-flow machines (i.e. where the fluid density is constant), dimensional analysis gives

$$\frac{P}{\rho N^3 D^5} = \phi \left(\frac{Q}{ND^3}, \frac{\rho ND^2}{\mu}, \frac{gH}{N^2 D^2} \right)$$

The various dimensionless groups can be combined to form another group, known as the specific speed, which can be based on the flow or shaft power:

$$N_{sp} = \left(\frac{N\sqrt{P/\rho}}{(gH)^{5/4}} \right), \quad N_{sQ} = \frac{N\sqrt{Q}}{(gH)^{3/4}}$$

The specific speed is an important parameter, in that the most efficient forms of a particular type of machine falls within a certain range of specific speed, and many of the machine design elements are functions of the specific speed. This is demonstrated by charts such as the Cordier diagram (Fig. 6.1).

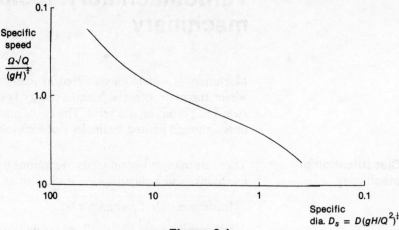

Figure 6.1

In machines where the density changes are significant the usual dimensionless groups are of the form

$$\frac{\Delta h_{0s}}{N^2 D^2}, \frac{P}{\rho_{01} N^3 D^5}, \eta = \phi\left(\frac{m}{\rho_{01} N D^3}, \rho_{01}\frac{ND^2}{\mu}, \frac{ND}{a_{01}}, \gamma\right)$$

$$\frac{p_{02}}{p_{01}}, \frac{\Delta T_0}{T_{01}}, \eta = \phi\left(\frac{m\sqrt{RT_{01}}}{p_{01} D^2}, \frac{ND}{\sqrt{RT_{01}}}, M, Re, \gamma\right)$$

Machine characteristics

Typical characteristics of various types of machines are shown in Fig. 6.2.

Efficient operation of a compressor is in the region to the right of the surge line. For multistage machines it commences at the point where, for a given $N/\sqrt{T_{02}}$, the pressure ratio is at about the maximum. This surge line gives the limit of *stable* operation, beyond which severe oscillation of the mass flow rate occurs.

Basic equations

In terms of two-dimensional flow the basic equations can be summarized as follows:

Continuity: $m = \rho A v_f$ = constant, where v_f = flow velocity
First law: $Q - W_s = \Delta h + \Delta(\tfrac{1}{2}v^2) + \Delta(gz)/\text{kg}$, or
                 $Q - W_s = \Delta h + \Delta(\tfrac{1}{2}v^2) = \Delta h_0$

where changes in potential energy are negligible

Second law: $T\,ds = du + p\,dv = dh - v\,dp$
Momentum: $W_s = m\Delta(uv_w)$ where v_w = whirl velocity

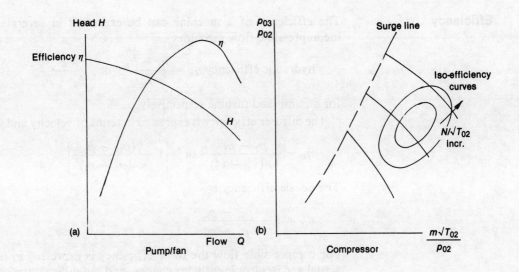
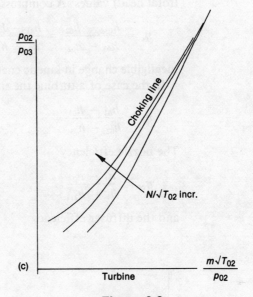

Figure 6.2

In the case of incompressible flow the density is constant, and temperature changes are negligible. Since enthalpy $h = u + pv$, then $\Delta h = \Delta(pv)$. The first law then takes the form of the Bernoulli equation:

$$\pm \rho E + p + \tfrac{1}{2}\rho v^2 + \rho g z + \Delta p_f = \text{constant, or}$$

$$h + \frac{v^2}{2g} + z + \Delta h_f \pm E = \text{constant}$$

The work done/kg is often termed the Euler head, E where $E = \Delta(uv_w)/g$.

Efficiency

The efficiency of a machine can be expressed in several forms. For incompressible-flow machines

$$\text{hydraulic efficiency } \eta_h = \frac{h_3 - h_2}{E} \quad \text{or} \quad \frac{E}{h_3 - h_2}$$

for a pump and turbine respectively.

The diffuser efficiency is expressed in terms of velocity and/or pressure.

$$\eta_D = \frac{p_2 - p_1}{\tfrac{1}{2}\rho(v_1^2 - v_2^2)} \quad \text{or} \quad \left[1 + \frac{(p_{01} - p_{02})}{(p_2 - p_1)}\right]^{-1}$$

The nozzle efficiency is

$$\eta_N = 1 - \frac{p_{01} - p_{02}}{p_1 - p_2}$$

For compressible flow the rotor efficiency is expressed in terms of the actual and isentropic enthalpy change, and can utilize static or stagnation (total head) values. A compressor isentropic efficiency is (Fig. 6.3)

$$\eta_c = \frac{h_{03s} - h_{02}}{h_{03} - h_{02}} = \frac{h_{3s} - h_2}{h_3 - h_2}$$

(negligible change in kinetic energy $\tfrac{1}{2}(v_3^2 - v_2^2)$).

In the case of a turbine the stagnation isentropic efficiency

$$\eta_t = \frac{h_{02} - h_{03}}{h_{02} - h_{03s}}$$

The nozzle efficiency

$$\eta_N = \frac{h_1 - h_2}{h_1 - h_{2s}}$$

and the diffuser efficiency

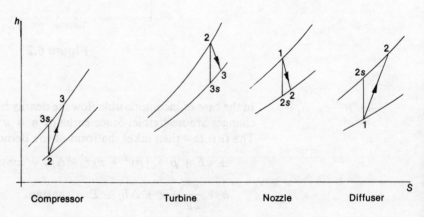

Figure 6.3

$$\eta_D = \frac{h_{2s} - h_1}{h_2 - h_1}$$

Small-stage (isentropic) efficiency The compression (or expansion in a machine) is often carried out in several stages. If the efficiency of each stage is the same, η_p, the efficiency of the whole machine η_c is not equal to η_p.

For an ideal gas

$$\eta_c = (R^{(\gamma-1)/\gamma} - 1)/(R^{(\gamma-1)/\eta_p \gamma} - 1) \qquad \text{(compression)}$$

$$\eta_t = (1 - R^{\eta_p(\gamma-1)/\gamma})/(1 - R^{(\gamma-1)/\gamma}) \qquad \text{(expansion)}$$

Steam turbines: reheat factor The ideal gas laws do not apply to steam, and the cumulative effect of friction in each stage is allowed for by the reheat factor. This is equal to $\Sigma \Delta h_s / (h_2 - h_{3s})$ (Fig. 6.4).

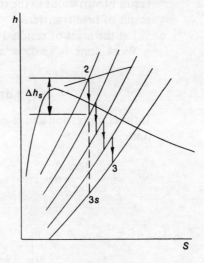

Figure 6.4

Slip factor: degree of reaction

In a centrifugal machine the fluid leaves the pump or compressor with imperfect guidance from the vanes, and the actual whirl velocity is less than the ideal velocity. The ratio of the actual to ideal whirl velocity is the slip factor.

Alternatively the effect is measured by means of a work done factor, which is the ratio of the actual to ideal work done.

In a machine (except pure impulse types) there is a pressure or enthalpy change across the stage. The stage normally consists of a fixed row of blades (stator, or guide vanes) and a moving row (rotor). The degree of reaction is

$$\Delta h (\text{rotor})/\Delta h (\text{stage}) \quad \text{or} \quad \Delta p (\text{rotor})/\Delta p (\text{stage})$$

(*Note:* In this chapter the notation used is outlined below. Unfortunately in the field of machinery there is a wide variety of notations, and this may need to be considered.

Subscripts: 0 stagnation (total head) conditions
1 stator, guide vane inlet
2 rotor inlet
3 rotor exit
4 diffuser, guide vane exit

Angles: blade and flow — measured in **all** cases from the tangential direction, in the direction of rotation.)

Reciprocating compressors

The operation of a reciprocating compressor is a mechanical cycle whose execution does not necessarily cause the working fluid to pass through a thermodynamic cycle. The pv diagram shown in Fig. 6.5 is not a thermodynamic cycle, and the work transfer should be evaluated from the steady-flow energy equation. The volume changes 2–3 and 4–1 are the result of variations in the mass of fluid contained in the cylinder, *not* as a result of heat transfer.

Let the mass of residual fluid = m kg and fluid delivered 1 kg.

Work done = $\oint p\,dv$ = area of loop 1234

$$= m \int_4^3 v\,dp - (m + 1) \int_1^2 v\,dp$$

If the compression and re-expansion laws are the same (reversibility) work done

$$= -\int_{p_1}^{p_2} v\,dp$$

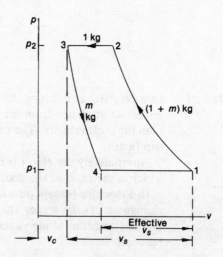

Figure 6.5

For polytropic compression

$$W = \frac{n}{n-1} p_1 v_1 \left[1 - \left(\frac{p_2}{p_1}\right)^{(n-1)/n} \right]$$

For isothermal compression

$$W = p_1 v_1 \ln(p_1/p_2)$$

Isothermal efficiency = isothermal work/actual work.
Volumetric efficiency = volume of fluid/cycle at NTP/swept volume.

Multi-stage compression

If the fluid is compressed in stages, with constant-pressure intercooling between stages, the ideal isothermal compression is approached. Fig. 6.6 shows that the saving in work input is increased as the number of stages increases.

If the intercooling between stages is complete, i.e. $T_1 = T_2' = T_3' = T_4'$, then for minimum total work, $p_1/p_2 = p_2/p_3 = p_3/p_4$. Under these conditions the work transfer in each stage is equal.

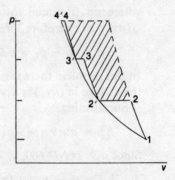

Figure 6.6

6.1 Centrifugal pump

A centrifugal pump impellor has an outer diameter of 30 cm, and discharge area = 0.10 m². The blades are backward curved at 145°, and there is no whirl at inlet. The flow velocity is constant, and the velocity leaving the diffuser = $K \times$ velocity leaving the impellor.

Show that, in general, the head developed is given by

$$H = aN^2 + bNQ + cQ^2$$

where N = rotational speed (rev/min), Q = flow rate, and a, b, c are functions of the speed and impellor dimensions.

Determine the values of a, b, c if $K = \frac{1}{2}$; and the hydraulic efficiency when $N = 1500$ rev/min, $Q = 0.2$ m³/s.

Solution The velocity triangles for the impellor are shown in Fig. 6.7.

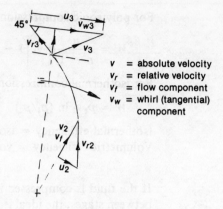

Figure 6.7

The flow is incompressible and applying the energy (Bernouilli) equation across the impellor

$$p_2 + \tfrac{1}{2}\rho v_2^2 + \rho E = p_3 + \tfrac{1}{2}\rho v_3^2$$

where no frictional losses are included, and E = work done/kg = $\Delta(uv_w)/g$. Therefore, pressure rise across the impellor is

$$\Delta p = p_3 - p_2 = \rho \Delta(uv_w) + \tfrac{1}{2}\rho(v_2^2 - v_3^2)$$

In this problem there is no whirl at the impellor inlet, i.e. $v_{w2} = 0$. Hence the change in (uv_w) is $u_3 v_{w3}$.

Substituting,

$$\Delta p = \rho u_3 v_{w3} + \tfrac{1}{2}\rho(v_2^2 - v_3^2)$$

Since $u_3 = \pi N D_3/60$, $v_f = Q/A_3$, $v_2 = v_f$,

$$v_{w3} = u_3 - v_f \cot(\pi - \theta_3) = u_3 + v_f \cot \theta_3,$$

the increase in head is

$$H = \frac{\Delta p}{\rho g}$$

Therefore, $H = \dfrac{u_3 v_{w3}}{g} + \dfrac{1}{2g}(v_f^2 - v_f^2 - v_{w3}^2) = \dfrac{u_3 v_{w3}}{g} - \dfrac{v_{w3}^2}{2g}$

$$= \frac{\pi N D_3}{60g}\left[\frac{\pi N D_3}{60} + \frac{Q}{A_3}\cot\theta_3\right]$$

$$- \frac{1}{2g}\left[\frac{\pi N D_3}{60} + \frac{Q}{A_3}\cot\theta_3\right]^2$$

There is also a pressure rise in the diffuser, in which the fluid enters at conditions 3, and leaves at conditions 4. The diffuser vanes are fixed and therefore there is no work done by or on the fluid. The energy equation gives, across the diffuser,

$$p_3 + \tfrac{1}{2}\rho v_3^2 = p_4 + \tfrac{1}{2}\rho v_4^2$$

therefore, $p_4 - p_3 = \tfrac{1}{2}\rho(v_4^2 - v_4^2) = \tfrac{1}{2}\rho v_3^2(1 - K^2)$

therefore, $H = \dfrac{p_4 - p_3}{\rho g} = \dfrac{v_3^2}{2g}(1 - K^2) = \dfrac{(1 - K^2)}{2g}[v_f^2 + v_{w3}^2]$

$$= \dfrac{(1 - K)^2}{2g}\left[\left(\dfrac{Q}{A_3}\right)^2 + \left(\dfrac{\pi N D_3}{60} + \dfrac{Q}{A_3}\cot\theta_3\right)^2\right]$$

The total increase in head is the sum of the two increases in the impellor and diffuser.

Therefore $H = aN^2 + bNQ + cQ^2$,

where $a = \left(\dfrac{K^2 - 2K + 2}{2g}\right)\left(\dfrac{\pi D_3}{60}\right)^2$

$b = \dfrac{(1 - K)^2}{2g}\left(\dfrac{\pi D_3}{60}\right)\dfrac{\cot\theta_3}{A_3}$

$c = \dfrac{1 + 2(1 - K)^2}{2gA_3^2}$

In this problem, $D_3 = 0.3$ m, $\theta_3 = 145°$, $A_3 = 0.10$ m^2 and $K = \tfrac{1}{2}$. Therefore, substituting these values gives

$a = 1.572 \times 10^{-5}, b = -0.00286, c = 7.645$

When $N = 1500$, $Q = 0.2$ substitution gives $H = 34.82$ m. Also

$E = \dfrac{u_3 v_{w3}}{g} = \dfrac{\pi(0.3)(1500)}{60g}\left[\dfrac{\pi(0.3)(1500)}{60} + \dfrac{0.2}{0.1}\cot 145\right]$

$= 2.402(23.56 - 2.856) = 49.73$ m

Therefore, hydraulic efficiency $\eta_h = H/E = \mathbf{0.70}$.

6.2 Pump characteristics

A pump, running at 2000 rev/min, gave the following results in a test:

Head H (m)	29.5	32.5	30.1	26.2	19.3	12.8
Flow Q (l/s)	0	6.62	11.05	15.75	22.05	27.0
Shaft power P (kW)	—	4.77	5.79	6.56	7.15	7.15

The pump delivers water against a resistance of $13.4 + 0.06Q^2$ m head.

Plot the pump characteristics and hence estimate the flow and power input at the operating point, and the efficiency and specific speed at this point.

Solution The determination of the power input and flow rate is obtained from the intersection of the pump characteristics and resistance curve (Fig. 6.8).

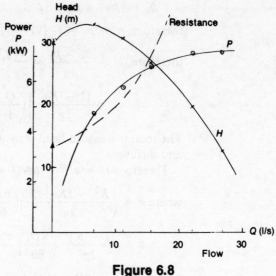

Figure 6.8

The equation of the latter curve is given as

$$H_R = 13.4 + 0.06 Q^2$$

From the graphs the operating conditions are

$$H = 26.9 \text{ m}, \, Q = 15 \text{ l/s}, \, P = 6.4 \text{ kW}$$

$$\text{Efficiency} = \frac{mgH}{P} = \rho g Q H / P$$
$$= 1000(9.81)(15 \times 10^{-3})(26.9)/6400$$
$$= \mathbf{0.62}$$

The specific speed $= \dfrac{N\sqrt{Q}}{(gH)^{3/4}} = \dfrac{2000 \times 2\pi/60 \times \sqrt{15 \times 10^{-3}}}{(9.81 \times 26.9)^{3/4}}$
$$= \mathbf{0.39}$$

at the operating point.

6.3 Hydraulic turbine

A Francis (inward flow radial) turbine operates under a net head of 80 m. Speed = 400 rev/min. Guide vane angle = 20°. The external diameter of the runner = 1.3 m. The radial velocity of flow is constant at 9 m/s. The water enters the draft tube, without whirl, at a velocity of 8 m/s and leaves at a velocity of 3 m/s.

The height of the runner entry (and guide vanes), and draft tube entry above the tailrace level is 2.5 m and 2.0 m respectively. The head loss in the guide vanes = 1.8 m, and in the draft tube = 3.0 m.

Calculate the pressure head at the runner inlet and exit, the head loss (due to friction) in the runner, and hydraulic efficiency.

Briefly outline the purpose of the draft tube.

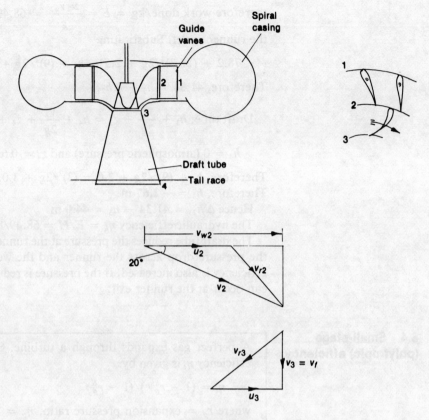

Figure 6.9

Solution Refer to Fig. 6.9. The energy (Bernoulli) equation is applied across the guide vanes, runner and draft tube.

Guide vanes: $h_1 + \dfrac{v_1^2}{2g} + z_1 = h_2 + \dfrac{v_2^2}{2g} + z_2 + \Delta h_{12}$

where Δh_{12} = head loss in friction. The guide vanes are fixed and there is no work done by the water. Hence, taking all static pressures above atmosphere (i.e. gauge)

$$80 + \dfrac{v_1^2}{2g} + 2.5 = h_2 + \dfrac{v_2^2}{2g} + 2.5 + 1.8$$

Also the velocity at the guide vane inlet v_1 and at the exit v_2 are equal, so that the pressure head at the runner inlet is $h_2 = 80 - 1.8 = \mathbf{78.2\ m}$.

Runner: $h_2 + \dfrac{v_2^2}{2g} + z_2 = h_3 + \dfrac{v_3^2}{2g} + z_3 + E + \Delta h_{23}$

From the velocity triangles $v_2 \sin 20 = v_f = 9$, therefore $v_2 = 26.3$ m/s, $v_{w2} = v_2 \cos 20 = 24.7$ m/s. Also $u_2 = \pi(1.3)(400)/60 = 27.2$ m/s, therefore work done/kg = $E = \dfrac{u_2 v_{w2}}{g}$ = 68.49 m. There is no whirl at the runner outlet. Substituting

$$78.2 + (26.3)^2/2g + 2.5 = h_3 + (9)^2/2g + 2.0 + 68.49 + \Delta h_{23}$$

Therefore, $41.34 = h_3 + \Delta h_{23}$.

Draft tube: $h_3 + \dfrac{v_3^2}{2g} + z_3 = h_4 + \dfrac{v_4^2}{2g} + z_4 + \Delta h_{34}$

$h_4 = 0$ (atmospheric pressure) and $z_4 = 0$ (datum)

Therefore, $h_3 + (9)^2/2g + 2.0 = (3)^2/2g + 3.0$.
Therefore, $h_3 = -2.67$ m.
Hence $\Delta h_{23} = 41.34 - h_3 =$ **44.0 m**
The hydraulic efficiency $\eta_h = E/H = 68.49/80 =$ **0.86**.

The draft tube reduces the pressure at the runner exit, thereby increasing the pressure drop across the runner and the work done. The hydraulic efficiency is also increased. If the pressure is reduced too much cavitation can occur at the runner exit.

6.4 Small-stage (polytropic) efficiency

A perfect gas expands through a turbine. Show that the overall efficiency η_t is given by

$$\eta_t = (1 - r_p^{k\eta_p})/(1 - r_p^K)$$

where r_p = expansion pressure ratio, η_p = small-stage efficiency and $K = (\gamma - 1)/\gamma$. Also show that $\eta_t > \eta_p$.

Derive the corresponding result in a compressor process.

Solution Referring to Fig. 6.10, the overall expansion from point 2 to point 3 is divided into several small stages 2–a, a–b, ...

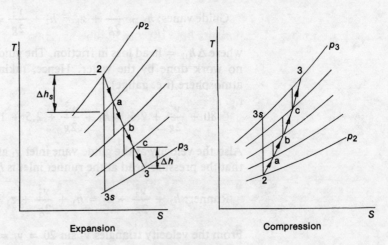

Figure 6.10

Consider one small stage $\eta_p = \dfrac{dh}{dh_s} = \dfrac{C_p\, dT}{v\, dp}$

since $T\, ds = dh - v\, dp$ and for an isentropic process $ds = 0$, therefore $dh_s = v\, dp$.

Also for an ideal gas $pv = RT$, $C_p = \dfrac{\gamma R}{\gamma - 1}$, $R = pv/T$.

Therefore, $\eta_p = \dfrac{\gamma R}{\gamma - 1} \dfrac{dT}{v\, dp} = \dfrac{\gamma}{\gamma - 1} \dfrac{pv}{T} \dfrac{dT}{v\, dp} = \dfrac{\gamma}{\gamma - 1} \dfrac{p}{T} \dfrac{dT}{dp}$

Therefore, $dT/T = \dfrac{\gamma - 1}{\gamma} \eta_p \dfrac{dp}{p}$

Integrating over the whole expansion, assuming that η_p is constant, gives

$$\int_{T_2}^{T_3} \dfrac{dT}{T} = \dfrac{\gamma - 1}{\gamma} \eta_p \int_{p_2}^{p_3} \dfrac{dp}{p}$$

Therefore, $\ln \dfrac{T_3}{T_2} = K\eta_p \ln \dfrac{p_3}{p_2}$

Therefore, $\dfrac{T_3}{T_2} = \left(\dfrac{p_3}{p_2}\right)^{K\eta_p} = r_p^{K\eta_p}$

In the ideal case, expansion from point 2 to point 3s,

$\dfrac{T_{3s}}{T_2} = r_p^{(\gamma - 1)/\gamma} = r_p^{K}$

Hence the overall efficiency is

$\eta_t = \dfrac{T_2 - T_3}{T_2 - T_{3s}} = \dfrac{1 - r_p^{K\eta_p}}{1 - r_p^{K}}$

Since this is an expansion process, $r_p = p_3/p_2 < 1$. Also the constant pressure lines *diverge* so that

$\dfrac{\Delta \delta h}{h_2 - h_{3s}} > \dfrac{dh}{dh_s}$

i.e. $\eta_t > \eta_p$.

In the case of a compression process

$\eta_p = \dfrac{dh_s}{dh} = \dfrac{v\, dp}{C_p\, dT} = \dfrac{\gamma - 1}{\gamma} \dfrac{T}{p} \dfrac{dp}{dT}$ and $\dfrac{dT}{T} = \dfrac{\gamma - 1}{\gamma} \dfrac{dp}{p}$

Integrating and rearranging gives

$\eta_c = \dfrac{r_p^{K} - 1}{r_p^{K/\eta_p} - 1}$

In this case $\eta_c < \eta_p$.

6.5 Steam turbine: reheat factor

A steam turbine has 5 stages. It is supplied with steam at 17 bar, 300 °C and exhausts at 0.04 bar, 88 % dry. The steam conditions at each stage exit are as shown in Table 6.1.

Table 6.1

stage exit number	1	2	3	4
pressure (bar)	7.5	2.75	0.8	0.2
condition	230 °C	140 °C	96 % dry	92 % dry

Using the Molier chart determine the internal (overall) efficiency of the turbine, the efficiency of each stage, the reheat factor and work output/kg.

Solution Referring to Fig. 6.11, the enthalpy of the steam at the various points is determined.

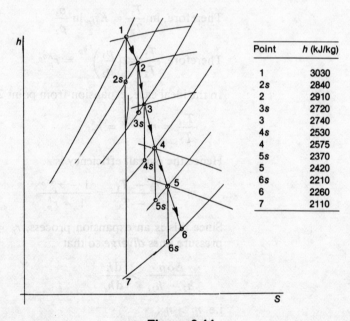

Point	h (kJ/kg)
1	3030
2s	2840
2	2910
3s	2720
3	2740
4s	2530
4	2575
5s	2370
5	2420
6s	2210
6	2260
7	2110

Figure 6.11

The overall efficiency is

$$\frac{h_1 - h_6}{h_1 - h_7} = (3030 - 2260)/(3030 - 2110) = \mathbf{0.837}$$

The stage efficiencies $\dfrac{h_1 - h_2}{h_1 - h_{2s}}$, $\dfrac{h_2 - h_3}{h_2 - h_{3s}}$, ... are

Stage	1	2	3	4	5
Efficiency	0.632	0.895	0.786	0.756	0.762

Reheat factor = $\Sigma \Delta h_s/(h_1 - h_7)$ = (190 + 190 + 210 + 205 + 210)/920
= **1.09**

Work output/kg = $\Sigma \Delta h$ = 120 + 170 + 165 + 155 + 160 = **790 kJ/kg**.

6.6 Steam turbine

A stage of an impulse steam turbine is velocity compounded with two rows of moving blades. The isentropic enthalpy drop for the stage is 320 kJ/kg. Nozzle angle = 16°, and velocity coefficient = 0.95. Blade speed = 150 m/s. Blade velocity coefficient (all blades) = 0.90. Steam flow rate = 20 kg/s. All blades are symmetrical.

Determine the blade angles, power output, stage efficiency and kinetic energy of the steam leaving the stage.

Solution The arrangement of the stage is shown in Fig. 6.12, and the velocity triangles in Fig. 6.13.

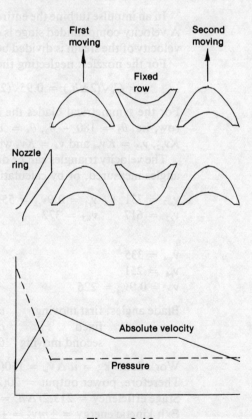

Figure 6.12

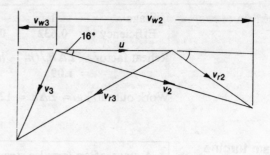

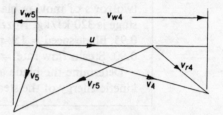

Figure 6.13

In an impulse turbine the entire pressure drop takes place in the nozzles. A velocity-compounded stage is one in which the total change in absolute velocity of the steam is divided between two or more moving (rotor) rows.

For the nozzles: neglecting the velocity at inlet

$$v_2 = C_v\sqrt{(2\Delta h_s)} = 0.95\sqrt{(2 \times 320 \times 1000)} = 760 \text{ m/s}$$

For the symmetrical blades the inlet and outlet angles are equal in each row, i.e. $\theta_2 = 180 - \theta_3$, $\theta_4 = 180 - \theta_5$. Also for impulse blades $v_{r3} = Kv_{r2}$, $v_{r5} = Kv_{r4}$ and $v_4 = Kv_3$ where K = blade velocity coefficient.

The velocity triangles can be drawn to scale and the velocities (m/s) and angles measured, or by calculation

$v_{w2} = 731$ $v_{r3} = 0.9v_{r2} = 555$ $v_3 = 417$
$v_{r2} = 617$ $v_{w3} = 372$ $v_4 = 0.9v_3 = 375$
  $v_{w5} = 17$

$v_{w4} = 335$
$v_{r4} = 251$
$v_{r5} = 0.9v_{r4} = 226$

Blade angles: first moving $\theta_2 = \mathbf{19.8°}$
 fixed $\alpha_4 = \mathbf{26.8°}$
 second moving $\theta_4 = \mathbf{42.4°}$

Work done/kg = $u\,\Delta v_w = 150(731 + 372 + 335 - 17) = 213.2$ kJ/kg.
Therefore, power output = $20(213.2) = \mathbf{4264}$ **kW**.
Stage efficiency = $213.2/\Delta h_s = 213.2/320 = \mathbf{0.667}$.
Exit kinetic energy = $\tfrac{1}{2}mv_5^2 = \tfrac{1}{2}(20)(153)^2\,W = \mathbf{234}$ **kW**.

6.7 Reaction steam turbine

> Show that in a reaction steam turbine stage the maximum stage efficiency is
>
> $$\frac{2\cos^2\alpha}{1+\cos^2\alpha}$$
>
> where α = nozzle (guide vane) angle. Assume that the reaction is 0.5, the axial flow velocity component is constant, and a normal stage.
>
> In a particular stage the mean diameter = 500 mm, blade height = 30 mm. The blade angles are 60° at inlet and 160° at outlet. Density of steam = 2.7 kg/m³. Speed = 3000 rev/min. Calculate the mass flow rate of steam, power developed and stage efficiency.

Solution The velocity triangles and stage diagrams are shown in Fig. 6.14.

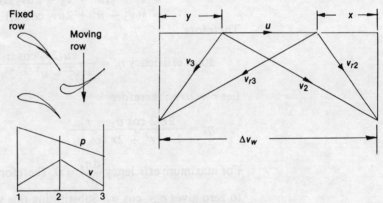

Figure 6.14

A normal stage is one in which the absolute velocity at inlet to and outlet from the stage is equal, i.e. $v_1 = v_3$.

Also the degree of reaction

$$\frac{h_2 - h_3}{h_1 - h_3} = 0.5$$

therefore, $h_2 - h_3 = h_1 - h_2 = 0.5(h_1 - h_3)$.

Stage efficiency = work done/energy input
$$= u(v_{w2} + v_{w2})/(h_1 - h_3 + \tfrac{1}{2}v_1^2).$$

Now no work is transferred from the steam *relative* to the moving blades, therefore the energy equation gives

$$h + \tfrac{1}{2}v_r^2 = \text{constant}$$

therefore, $h_2 - h_3 = \tfrac{1}{2}(v_{r3}^2 - v_{r2}^2)$.

The corresponding equation for the fixed guide blades is

$$h_1 - h_2 = \tfrac{1}{2}(v_2^2 - v_1^2)$$

Referring to the velocity triangles, and using the reaction = 0.5 gives
$$h_1 - h_2 = h_2 - h_3.$$
Therefore $v_2^2 - v_1^2 = v_{r3}^2 - v_{r2}^2 = v_2^2 - v_3^2.$
Therefore, $(u + v)^2 + v_f^2 - x^2 - v_f^2 = (u + x)^2 + v_f^2 - y^2 - v_f^2$
Therefore, $(u + y)^2 - x^2 = (u + x)^2 - y^2$
Therefore, $2uy + y^2 - x^2 = 2ux + x^2 - y^2$
Therefore, $2(y^2 - x^2) = 2u(x - y)$
Therefore, $y + x = u$ or $x = y$.

Hence for 0.5 reaction the velocity triangles are symmetrical (but **not** the blades), i.e. the fixed and moving blades are of the same shape.

Work done $= u(v_{w2} + v_{w3}) = u(v_2 \cos \alpha_2 + y) = u(v_2 \cos \alpha_2 + x)$
$= u(v_2 \cos \alpha_2 + v_2 \cos \alpha_2 - u) = u(2v_2 \cos \alpha_2 - u)$

Energy input $= h_1 - h_3 + \tfrac{1}{2}v_1^2 = 2(h_1 - h_2) + \tfrac{1}{2}v_1^2 = (v_2^2 - v_1^2)$
$+ \tfrac{1}{2}v_1^2 = v_1^2 - \tfrac{1}{2}v_1^2 = v_2^2 - \tfrac{1}{2}v_3^2 = v_2^2 - \tfrac{1}{2}v_{r2}^2$
$= v_2^2 - \tfrac{1}{2}(u^2 + v_2^2 - 2uv_2 \cos \alpha_2)$
$= \tfrac{1}{2}(v_2^2 - u^2 + 2uv_2 \cos \alpha_2)$

Therefore,
$$\text{stage efficiency } \eta_s = \frac{u(2v_2 \cos \alpha_2 - u)}{\tfrac{1}{2}(v_2^2 - u^2 + 2uv_2 \cos \alpha_2)}$$

Let $r = u/v_2$; therefore
$$\eta_s = \frac{2r(2 \cos \alpha_2 - r)}{1 - r^2 + 2r \cos \alpha_2}$$

For maximum efficiency $\dfrac{d\eta_s}{dr} = 0$. Therefore differentiating and equating to zero gives $r = \cos \alpha_2$. Substituting this value of r gives the maximum stage efficiency as

$$\eta_s \text{ (max.)} = \frac{2 \cos^2 \alpha_2}{1 + \cos^2 \alpha_2}$$

In this particular stage $u = \pi(0.5)(3000)/60 = 78.54$ m/s. The velocity triangles are shown in Fig. 6.15.

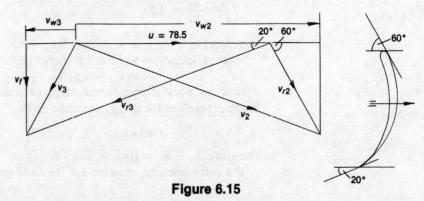

Figure 6.15

$v_f = 36.2$, $v_{w2} = 99.5$, $v_{w3} = 21.0$, $v_2 = 105.9$, $v_3 = 41.9$ m/s, therefore,
mass flow rate = $\rho A v_f = 2.7 \times \pi (0.5)(0.03)(36.2)$
= **4.606 kg/s**
Power output = $mu(v_{w2} + v_{w3})$ = 43 570 W or **43.57 kW**
Energy input = $m(v^2 - \tfrac{1}{2} v_3^2) = m[(105.9)^2 - \tfrac{1}{2}(41.9)^2]$ = 47 610 W
Therefore, η_s = 43.57/47.61 = **0.915**.

6.8 Centrifugal compressor

A centrifugal compressor draws in air at 120 m/s, and static conditions of 0.90 bar, 7 °C. The air is compressed to static conditions of 1.40 bar, 70 °C and leaves with an absolute velocity of 300 m/s. Flow rate = 3 kg/s.
C_p = 1.005 kJ/kg K. Assuming that there is no whirl at inlet calculate the compressor drive power and static isentropic efficiency.
A diffuser is fitted to the impellor outlet, reducing the air velocity to 150 m/s. Diffuser efficiency = 70 %. Calculate the pressure at the diffuser outlet and overall static isentropic efficiency.

Solution Referring to Fig. 6.16, the energy equation applied to the impellor gives

$$h_2 + \tfrac{1}{2} v_2^2 + E = h_3 + \tfrac{1}{2} v_3^2$$

where E = work done/kg = $u_3 v_{w3}$ (no whirl at inlet). Hence

$$E = h_3 - h_2 + \tfrac{1}{2}(v_3^2 - v_2^2)$$
$$= C_p(T_3 - T_2) + \tfrac{1}{2}(v_3^2 - v_2^2)$$
$$= 1.005(70 - 7) + \tfrac{1}{2}(300^2 - 120^2) \times 10^{-3} = 101 \text{ kJ/kg}$$

Therefore, power input = 3(101) = **303 kW**.
If the compression was isentropic

$$T_{3s} = T_2(p_3/p_2)^{\gamma-1/\gamma} = 280(1.4/0.9)^{0.2857} = 318 \text{ K}$$

Therefore, isentropic work

$$= 1.005(318 - 280) + \tfrac{1}{2}(300^2 - 120^2) \times 10^{-3} = 76.0 \text{ kJ/kg}$$

Therefore, static isentropic efficiency = 76.0/101 = **0.752**.
(*Note:* The isentropic efficiency can also be determined in terms of the stagnation temperature.

$$T_{02} = T_2 + v_2^2/2C_p = 280 + (120)^2/2010 = 287 \text{ K}$$
$$T_{03} = T_3 + v_3^2/2C_p = 343 + (300)^2/2010 = 388 \text{ K}$$
$$T_{03s} = T_{3s} + v_3^2/2C_p = 318 + (300)^2/2010 = 363 \text{ K}$$

Therefore stagnation isentropic efficiency = (363 − 287)/(388 + 287)
= 0.752.)

For the diffuser

$$\eta_D = \frac{T_{4s} - T_3}{T_4 - T_3}$$

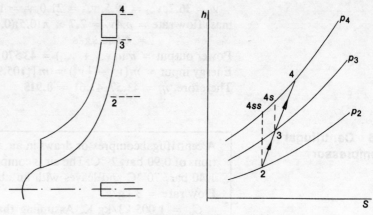

Figure 6.16

therefore, $0.7(T_4 - 343) = T_{4s} - 343$
and since no work is done in the diffuser, $T_{04} = T_{03} = 388$ K.
Therefore, $T_4 = 388 - v_4^2/2C_p = 388 - (150)^2/2010 = 377$ K
Therefore, $T_{4s} = 0.7(377 - 343) + 343 = 367$ K.

Now $\dfrac{T_{4s}}{T_3} = \left(\dfrac{p_4}{p_3}\right)^{(\gamma-1)/\gamma}$

Therefore, $\dfrac{p_4}{p_3} = \left(\dfrac{367}{343}\right)^{3.50} = 1.267$

Therefore, $p_4 = $ **1.774 bar**.

The overall isentropic temperature rise is $T_{4ss} - T_2$ where

$\dfrac{T_{4ss}}{T_2} = \left(\dfrac{p_4}{p_2}\right)^{\gamma/(\gamma-1)} = \left(\dfrac{1.774}{0.9}\right)^{0.2857} = 1.214$

Therefore, $T_{4ss} = 340$ K.
Therefore, overall static isentropic efficiency

$= (340 - 280)/(377 - 280)$
$= \mathbf{0.618}.$

6.9 Axial flow compressor

An axial flow compressor has a degree of reaction = 0.5. Show that the pressure ratio is given by

$$\dfrac{p_3}{p_1} = \left(\dfrac{\eta_s u \lambda \Delta v_w}{C_p T_1} + 1\right)^{\gamma/(\gamma-1)}$$

where λ = work done factor.

In a certain stage the mean diameter = 0.85 m, and speed = 5300 rev/min. Isentropic efficiency = 82 %. Pressure ratio = 1.40. Air inlet temperature = 24 °C. Flow velocity (constant) = 160 m/s. Determine the blade angles at inlet and outlet.

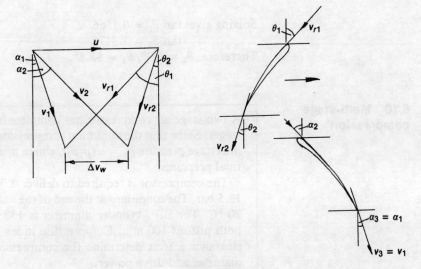

Figure 6.17

Solution The velocity triangles are shown in Fig. 6.17. The usual convention in this field is to measure the angles from the axial direction, and is adopted in the solution.

For a normal stage $v_3 = v_1$, and for 0.5 reaction the velocity triangles are symmetrical.

Hence

$$\text{work done/kg, } E = h_3 - h_1 + \tfrac{1}{2}(v_3^2 - v_1^2)$$
$$= h_3 - h_1$$
$$= C_p(T_3 - T_1)$$
$$= C_p(T_{3s} - T_1)/\eta_s$$

But $\dfrac{T_{3s}}{T_1} = \left(\dfrac{p_3}{p_1}\right)^{(\gamma-1)/\gamma}$, and also $E = \lambda u \Delta v_w$.

Therefore, $\lambda u \Delta v_w = \dfrac{C_p T_1}{\eta_s}\left[\left(\dfrac{p_3}{p_1}\right)^{(\gamma-1)/\gamma} - 1\right]$

and rearranging $\dfrac{p_3}{p_1} = \left(\dfrac{\eta_s \lambda u \Delta v_w}{C_p T_1}\right)^{\gamma/(\gamma-1)}$

In this given stage, $u = \pi(0.85)(5300)/60 = 235.9$ m/s, and substituting

$$\dfrac{p_3}{p_1} = 1.4 = \left(\dfrac{0.82 \times 0.85 \times 235.9 \,\Delta v_w}{1005 \times 301} + 1\right)^{3.5}$$

Therefore, $1.101 = 5.435 \times 10^{-4}\,\Delta v_w + 1$
Therefore, $\Delta v_w = 185.8$ m/s.

Referring to the velocity triangles $\Delta v_w = v_f(\tan\theta_2 - \tan\theta_1)$ therefore, $\tan\theta_2 - \tan\theta_1 = 185.8/160 = 1.1613$, and

$$u = v_f(\tan\theta_1 + \tan\theta_2)$$

therefore, $\tan\theta_1 + \tan\theta_2 = 235.9/160 = 1.4744$

Solving gives $\tan \theta_1 = 0.1566$
$\tan \theta_2 = 1.3179$
Therefore, $\theta_1 = 8.9°$, $\theta_2 = 52.8°$.

6.10 Multi-stage compression

A two-stage air compressor has complete intercooling between the stages. Show that the minimum compression work is obtained if the interstage pressure $p_2 = \sqrt{(p_1 p_3)}$, where p_1 and p_3 are the initial and final pressures.

The compressor is required to deliver 0.008 kg/s at a pressure of 13.5 bar. The conditions at the end of the suction stroke are 0.95 bar, 20 °C. The L.P. cylinder diameter is 150 mm, and the stroke of both pistons 100 mm. Compression index = 1.20. Neglecting the clearance effects determine the compressor speed, H.P. cylinder diameter and drive power.

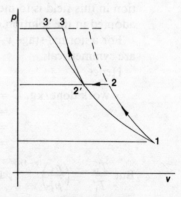

Figure 6.18

Solution Referring to Fig. 6.18, the total compression work is

$$W_c = \frac{n}{n-1} p_1 v_1 [1 - (p_2/p_1)^{(n-1)/n}]$$

$$+ \frac{n}{n-1} p_2 v_2 [1 - (p_3/p_2)^{(n-1)/n}]$$

For complete intercooling $T'_2 = T_2$, therefore $p_2 v'_2 = p_1 v_1$, therefore

$$W_c = \frac{n}{n-1} p_1 v_1 [2 - (p_2/p_1)^{(n-1)/n} - (p_3/p_2)^{(n-1)/n}]$$

The only variable is the interstage pressure p_2, therefore for minimum work $dW_c/dp_2 = 0$.

Therefore, $-\left(\dfrac{n-1}{n}\right)(p_2/p_1)^{(n-1)/n-1} - \left(\dfrac{n-1}{n}\right)(p_3/p_2)^{(n-1)/n-1} = 0$

Therefore, $p_2/p_1 = p_3/p_2$ or $p_2 = \sqrt{(p_1 p_3)}$

Since the pressure ratio is the same stage the compression work in each stage is the same.

In this problem $p_2 = 0.95 \times 13.5 = 3.58$ bar
Volume of air drawn in $= mRT/p = 0.008 \times 287 \times 293/(0.95 \times 10^5)$
$= 0.00708$ m^3/s

Volume drawn in/cycle $= 0.1 \times \dfrac{\pi}{4}(0.15)^2 = 0.00177$ m^3

Therefore, speed $= 0.00708/0.00177 = 4.006$ rev/s $= $ **240 rev/min**.
For the H.P. cylinder $p_2' v_2'/T_2' = p_2 v_2/T_2$ and $T_2' = T_2$
Therefore $v_2' = 0.00177 \times 0.95/3.58 = 0.00047$ m^3

Therefore, H.P. diameter $= \dfrac{0.00047 \times 4}{\pi \times 0.1} = 0.077$ or **77 mm**.

(*Note*: The volume of air leaving the intercooler must equal the swept volume (effective) of the H.P. cylinder in the same time interval. If this was not the case the interstage pressure would automatically adjust itself to achieve this requirement.)

Compression work $W_c = \dfrac{2n}{n-1} p_1 v_1 [1 - (p_2/p_1)^{(n-1)/n}]$

$= \dfrac{2(1.20)}{0.20} \times 0.95 \times 10^5 \times 0.00708$

$\times [1 - (3.58/0.95)^{1/6}]$ W

$= $ **2.0 kW**.

6.11 Reciprocating compressor

A two-stage double acting air compressor has the following dimensions:
  L.P. cylinder: 300 mm diameter, 460 mm stroke,
    clearance 4 % of the swept volume
  H.P. cylinder: 180 mm diameter, 460 mm stroke,
    clearance 6 % of the swept volume
The ambient conditions are 25 °C, 1 bar. The pressure in the L.P. cylinder during suction is 0.95 bar, and the temperature 40 °C. Delivery pressure $= 7.0$ bar. The air temperature leaving the intercooler is 50 °C.
  Compression and re-expansion index is 1.30.
  Determine the pressure in the intercooler, volumetric efficiency and free air delivery at 360 rev/min.

Solution Referring to Fig. 6.19, the volume of cooled air/cycle leaving the intercooler = effective swept volume of the H.P. cylinder.

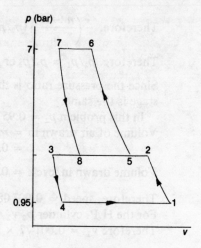

Figure 6.19

For the L.P. cylinder,

$$\text{swept volume} = \frac{\pi}{4}(0.3)^2(0.46) = 0.0325 \text{ m}^3$$

$$\text{clearance volume} = 0.04 v_s = 0.0013 \text{ m}^3$$

For the H.P. cylinder,

$$\text{swept volume} = \frac{\pi}{4}(0.18)^2(0.46) = 0.0117 \text{ m}^3$$

$$\text{clearance volume} = 0.06 \times 0.0117 = 0.0007 \text{ m}^3$$

Volume of air into L.P. cylinder/cycle

$$= v_1 - v_4$$
$$= (0.0325 + 0.0013) - 0.0013 \times (p/0.95)^{1/1.3}$$
$$= 0.0338 - 0.00125 p^{0.769} \text{ m}^3$$

where p = interstage pressure (bar). It should be noted that the clearance air at point 3 expands to point 4 as the pressure decreases from p bar to 0.95 bar.

Volume of air *leaving* intercooler at 50 °C is

$$(0.0338 - 0.00125 p^{0.769}) \times \frac{0.95}{p} \times \frac{323}{313}$$

$$= \frac{0.98}{p}(0.0338 - 0.00125 p^{0.769}) \text{ m}^3$$

The swept volume of the H.P. cylinder is

$$v_5 - v_8 = (0.0117 + 0.007) - 0.0007\left(\frac{7.0}{p}\right)^{0.769}$$

$$= 0.0124 - 0.00313/p^{0.769}$$

Since these two volumes must be the same

$$\frac{0.98}{p}(0.0338 - 0.00125p^{0.769}) = 0.0124 - 0.00313/p^{0.769}$$

and solving gives $p = $ **2.79 bar**.

Air volume drawn in/cycle

$$= 0.0338 - 0.00125(2.79)^{0.769} = 0.0310 \text{ m}^3$$

Therefore, free air volume (ambient conditions)

$$= 0.0310 \times \frac{0.95}{1} \times \frac{298}{313} = 0.028 \text{ m}^3$$

therefore, free air delivery at 360 rev/min

$$= 0.028 \times 360 \times 2 \text{ (double acting)} = \textbf{20.16 m}^3\textbf{/min}$$

Volumetric efficiency $= 0.028/0.0325 = $ **0.86**.

Problems

1 The characteristics of a fan are as shown in Table 6.2.

Table 6.2

Flow (m^3/s)	0	0.56	1.11	1.67	2.22	2.78	3.33
Fan pressure (N/m^2)	490	535	550	535	490	417	314
Fan input power (kW)	—	0.63	0.90	1.20	1.53	1.70	1.75

The fan delivers air along a duct, for which the resistance is $160Q^2$ N/m^2, where $Q = $ flow (m^3/s).

Determine the operating conditions and fan power.

Answer 525 N/m^2, 1.82 m^3/s; 1.28 kW

2 A pump has the characteristic shown in Table 6.3.

Table 6.3

Q (m^3/s)	0	0.225	0.335	0.425	0.545	0.650	0.750
H (m)	20	17	15	13	10	7	3
η (%)	—	43.5	54.5	59.8	56.7	49.0	39.5

The pump delivers water through a pipeline, and the system resistance is $5 + 44Q^2$ m, where $Q = $ flow rate (m^3/s).

Determine the operating point and pump power required.

Answer 13 m, 0.425 m^3/s; 90.6 kW

3 Water is pumped at a rate of 60 l/s through a pipeline, resistance $40 + 2.4Q^2/D^5$ m head. Q = flow rate (m³/s), D = pipe diameter (m).

The pump efficiency = 65%, and it is electrically driven at a cost of 3 p/kWh.

Maintenance cost = 10% capital cost/year.
Capital cost of pipe = £140/tonne.
Pipe length = 1200 m.

Table 6.4 shows how the mass per unit length varies with diameter.

Table 6.4

Pipe diameter (cm)	15	20	25	30	40
Mass per unit length (kg/m)	50	70	95	122	190

Determine the most economical pipe diameter based on a 6000 h working period/year, and a life of 10 years.
Answer 25 cm diameter

4 A centrifugal fan supplies 2 m³/s of air, at 1000 rev/min. The impellor diameter = 60 cm at outlet, 45 cm at inlet; and the width is constant at 15 cm.

The blades are curved with a blade angle at outlet = 30°. 40% of the absolute velocity head at the runner outlet is recovered in a diffuser.

Assuming that there is no whirl at inlet, and the air density is constant at 1.2 kg/m³, calculate the power input, pressure rise and efficiency when the blades are (a) backward curved, (b) forward curved. The pressure at inlet = 0.95 bar.
Answer (a) 1.73 kW, 625 N/m², 86.0%; (b) 3.95 kW, 996 N/m², 60.5%

5 An inward flow reaction hydraulic turbine runs at 480 rev/min.

	inlet	outlet
Runner diameter (cm)	90	50
Flow area (m²)	0.48	0.63

Flow rate = 4.3 m³/s. Guide vane exit angle = 40°.

Assuming that there are no frictional losses, no whirl at exit, and the pressure at the runner exit = 1 bar, calculate the head at the runner inlet, hydraulic efficiency, specific speed (based on power), and power output.
Answer 34.4 m, 82%, 0.012, 2.4 MW

6 An axial flow pump has a rotor with fixed blade angles, 30° at inlet and 45° at outlet, from the tangential direction. Guide vanes at the inlet side remove the inlet whirl. The flow velocity is constant at 4 m/s.

(a) Determine the pressure rise and hydraulic efficiency, and blade speed, when passing water.

(b) Determine the pressure rise and hydraulic efficiency if the blade speed is doubled.

Answer (a) 16 kN/m², 78.8 %, 6.93 m/s; (b) 16 kN/m², 39.4 %

7 Steam is supplied to a five-stage pressure-compounded turbine at 15 bar, 350 °C and leaves the last stage at 0.07 bar, 95 % dry.

The work output from each stage is equal, and the condition line is straight on the Molier chart.

Determine the overall efficiency, steam conditions at each stage exit, stage efficiency and reheat factor.

Answer 73.9 %, 6.2 bar, 274 °C: 2.4 bar, 200 °C: 0.85 bar, 125 °C; 0.26 bar, 99 % dry; 61.8, 65.3, 69.5, 71.6, 74.3 %; 1.08

8 In an impulse steam-turbine velocity-compounded stage, the mean diameter = 90 cm and the speed = 3000 rev/min. Steam is supplied to the nozzles at 20 bar, 250 °C and expands to a rotor chamber pressure of 8 bar. Nozzle angle = 18°. The steam velocity leaving the nozzle = 590 m/s.

Velocity coefficient (*all* blades) = 0.86.

The blade angles at exit are:

first moving   25°
fixed   28°
second moving 30°

Determine the blade efficiency, inlet blade angles, and steam condition at the stage outlet.

Answer 76 %; 23.4°, 37.7°, 59°; dry saturated

9 A single-stage, single-row impulse steam turbine has a nozzle angle of 20°, and the rotor angles are 30° (inlet) and 150° (exit) measured from the tangent in the direction of rotation. Nozzle exit velocity = 1200 m/s. Blade velocity coefficient = 0.85.

The steam is supplied at 10 bar, 350 °C and the exhaust pressure is 0.15 bar.

Calculate the work done/kg, blade efficiency and condition of the steam at exhaust.

Answer 549 kJ/kg, 76 %, 97 % dry

10 Two stages of a steam turbine consist of a simple impulse stage followed by a reaction stage.

The steam is expanded from 10 bar, 250 °C to $3\frac{1}{2}$ bar, dry saturated through a nozzle, angle 20°. The mass flow rate = 22 kg/s.

First stage: blade speed = 160 m/s
                outlet blade angle = 30° to the tangent
                blade velocity coefficient = 0.8
Second stage: blade speed = 220 m/s

stator row, outlet angle = 20°
jet speed = 440 m/s
rotor blade outlet angle = 18°
static enthalpy drop = 200 kJ/kg

Determine the nozzle efficiency, power output from each stage and stage efficiencies.

Answer 97 %; 2820 kW, 59 %; 3178 kW, 72 %

11 A steam turbine consists of 12 successive 50 % reaction stages, and runs at 3000 rev/min.

For each stage: mean diameter = 1.2 m
blade inlet angle = 80°
blade outlet angle = 20°

The axial flow velocity is constant, the total isentropic enthalpy drop through the turbine is 670 kJ/kg, and the reheat factor = 1.04.

Determine the enthalpy drop/stage, and stage efficiency.

Given that the turbine develops 75 MW estimate the blade height at the outlet from each stage.

The Molier diagram should be used for the solution. The steam enters the first stage at 150 bar, 400 °C.

Answer 58 kJ/kg, 70 %. Approx. 10, 12, 17, 22, 28, 38, 51, 70, 96, 140, 190, 245 mm

12 In an axial-flow steam-turbine stage the axial-flow velocity is constant. Speed = 3000 rev/min.

Blade height = 10 cm.
Mean blade diameter = 1 m.
50 % reaction at the mean blade height.
Absolute velocity of steam, stage inlet = 100 m/s.
Absolute velocity of steam, stage exit = 100 m/s.
Stage isentropic efficiency = 90 %.
Steam conditions at inlet = 7 bar, 300 °C.
Steam pressure at stage exit = 6 bar.

Assuming free vortex flow (i.e. work done/kg = constant at all radii) calculate the degree of reaction at the root and tip; blade angles at inlet and outlet, at the root, mean and tip diameters.

Answer 37, 58 %; root 51, 27°; mean 69, 25°; tip 88, 24°

13 A centrifugal compressor handling air runs at 3000 rev/min and develops a stagnation (total) pressure ratio of $3\frac{1}{2}$:1. Mass flow rate = 1.5 kg/s. The ambient conditions are 15 °C, 1 bar. Stagnation isentropic efficiency = 80 %. There is no whirl at inlet, and the impellor vanes are straight radial at the tip. Slip factor = 0.85. C_p = 1.005 kJ/kg K. Flow velocity = 150 m/s.

Calculate the impellor tip diameter and input power, static pressure at inlet and tip, and width at the tip.

Answer 27.3 cm, 235 kW; 0.87, 2.02 bar; 0.59 cm

14 A single-stage centrifugal supercharger is required to maintain a stagnation pressure of 1.5 bar in the manifold of an I.C. engine. The ambient suction conditions are -7 °C, 0.6 bar. The vanes are straight radial at the tip. Flow rate = 0.9 kg/s.

For the mixture $R = 275$ J/kg K, $\gamma = 4/3$
The stagnation isentropic efficiency = 75 %
Speed = 20 000 rev/min
Mixture velocity at entry = 100 m/s.

Calculate the impellor tip diameter, entry annulus area, and power input.

Answer 30.4 cm, 0.012 m^2, 91 kW

15 A turbo compressor delivers 11.7 m^3/s (measured at 0 °C, 1 bar). The air pressure is 1 bar at inlet, 7 bar at outlet. The air temperature = 15 °C at inlet, 95 °C at outlet. The heat extracted from the air = 2390 kW, and the power input = 3800 kW. $C_p = 1.005$ kJ/kg K.

Calculate the isothermal efficiency, and show that there is more than one stage.

Answer 63 %

16 A radial flow air compressor runs at 18 000 rev/min. The vanes are radial at outlet and the outer diameter is 50 cm. The stagnation conditions at inlet are 1 bar, 290 K. The air velocity entering the compressor and leaving the diffuser are equal, and there is no whirl at inlet. The flow velocity is constant at 150 m/s. $C_p = 1.005$ kJ/kg K. The static efficiency of the impellor + diffuser = 80 %.

Calculate the work done/kg, conditions at the diffuser exit.

Answer 222 kJ/kg; 2.63 bar, 500 K

17 A centrifugal compressor has the following dimensions:
Tip radius = 30 cm
Eye radius = 15 cm (tip), 5 cm (hub)
Stagnation conditions at inlet = 1 bar, 290 K
Overall stagnation pressure ratio = 4:1
Overall stagnation isentropic efficiency = 80 %
Slip factor = 0.85
Radial blades at the impellor tip.

The work done/kg is the same, at any radius, and the flow velocity = 100 m/s. $C_p = 1.005$ kJ/kg K. $\gamma = 1.40$. There is no whirl at inlet.

Calculate the rotational speed and the blade angle at the eye hub and tip.

Answer 14 600 rev/min; 110.9°, 138.9°

18 A centrifugal blower takes in air at 1 bar, 15 °C and delivers it at 1.4 bar.
Impellor diameter = 45 cm (inlet), 110 cm (outlet)
Speed = 5000 rev/min
Radial flow velocity is constant = 60 m/s

Isentropic efficiency = 80 %
Mass flow rate = 18 kg/s.

Assuming that there is no pre-whirl and no diffuser, the air density is constant at 1.20 kg/m³, and C_p = 1.005 kJ/kg K, calculate the vane angles and width at inlet and outlet, the power input and manometric (pressure) efficiency.

Answer 153°, 149.8°; 17.7, 7.2 cm; 959 kW; 62.6 %

19 A stage of an axial flow compressor has a mean blade diameter of 50 cm, and the air enters axially at stagnation conditions of 1 bar, 290 K.

At the mean diameter the rotor blade angles at entry and exit are 30° and 60° respectively (measured from the tangential direction).
Stagnation isentropic efficiency = 85 %
Work done factor = 0.86
Speed = 15 000 rev/min.

Assuming that the axial air velocity is constant, calculate the degree of reaction, stage stagnation pressure ratio, and static pressure and temperature at the rotor inlet and outlet.

C_p = 1.005 kJ/kg K. γ = 1.40.

Answer $\frac{2}{3}$; 2.24; 0.75 bar, 264 K; 1.22 bar, 318 K

20 An axial flow compressor has 8 stages, and runs at 9000 rev/min. The overall stagnation pressure ratio = 4:1. Mass flow rate = 16 kg/s. Overall stagnation isentropic efficiency = 86 %. Work done factor = 0.85. The blade speed and reaction at the mean diameter (all stages) = 200 m/s and 50 % respectively. The axial flow velocity is constant at 150 m/s.

The inlet stagnation conditions are 1 bar, 275 K.

Assuming that the work done is equally divided between the stages, determine the hub and tip diameters at the first stage and rotor blade angles at the mean diameter in the first stage.

Given that the work done/kg is constant over the blade height calculate the degree of reaction at the hub and tip.

Answer 460, 390 mm; 43.6, 74.1°; 0.49

21 Show that for an axial flow compressor the degree of reaction, R, is given by

$$R = 1 - (v_{w2} + v_{w3})/2u$$

where u = blade velocity
v_{w2}, v_{w3} = whirl velocity at the rotor inlet and exit

A compressor has 11 stages, and the stagnation conditions of the air at inlet are 1.0 bar, 290 K. The air stagnation pressure at outlet is 5.0 bar.

Stagnation isentropic efficiency = 79 %. Work done factor = 0.86. Mean blade speed = 150 m/s. Axial flow velocity = 135 m/s.

Degree of reaction = 60 %. C_p = 1.005 J/kg K.

Assuming that the work done is shared equally between the stages, calculate the stagnation temperature rise/stage, whirl velocity at inlet and exit, and the blade angles at the rotor inlet and exit.

Given that the speed = 6000 rev/min, mass flow rate = 20 kg/s, calculate the blade height at inlet to the first stage.

Answer $19\frac{1}{2}$ K; 136, −16 m/s; 39°, 84°, 8.9 cm

22 The rotor row of an axial compressor stage is such that the blade angle at outlet is constant over the whole blade height at 45°.

The axial velocity of flow (equal at inlet and exit) varies linearly from 60 m/s at the hub diameter of 76 cm to 120 m/s at the tip diameter of 106 cm. Speed = 5000 rev/min.

Assuming that there is no whirl at inlet and the air density is constant at 1.30 kg/m³, calculate the torque on the rotor, and hence the mean work done/kg over the blade height.

Answer 3580 Nm, 36.7 kJ/kg

23 Outline the importance of the term specific speed, and its use in the selection of turbo machinery types. Assuming the dimensional equation

$$\frac{P}{\rho N^3 D^5} = \phi \left[\frac{Q}{ND^3}, \frac{gH}{N^2 D^2}, \frac{\mu}{\rho ND^2} \right]$$

show that a specific speed $N_s = N\sqrt{Q}/(gH)^{3/4}$.

Select from the list in Table 6.5 a suitable type of machine for the duty stated:

(a) Pump to deliver 280 l/s water against a head of 9 m, at a speed of 1000 rev/min.

(b) An air compressor to deliver 15 m³/s against a pressure rise of 0.4 bar, at a speed of 5000 rev/min. Assume ρ = 1.20 kg/m³.

(c) Pump to deliver 0.5 m³/s water against a head of 12 m at a speed of 1500 rev/min.

Table 6.5

Type	N_s (rad/s)
Francis turbine	0.3−2.0
Kaplan turbine	2−5
Centrifugal pump	0.2−2.5
Axial-flow pump	2.5−5.5
Centrifugal compressor	0.5−2.0
Axial-flow compressor	1.5−2.0

Answer centr., centr., axial-flow

24 A reaction turbine is supplied with steam at 20 bar, 300 °C and exhausts at 0.1 bar. Assuming that the stage efficiency = 0.76 and the reheat factor = 1.035 estimate the steam consumption for a power output of 25 MW.

The turbine speed is 1500 rev/min. Calculate the blade length and mean diameter of the blade row at a point in the turbine where the pressure is 0.8 bar. Also estimate the number of stages required, if the work is divided equally between them. The stator and rotor blades are of the same section with an outlet angle of 20°, the axial velocity of flow is constant and equal to 0.5 of the blade velocity, and the blade length is one twelfth of the mean diameter of the blade row at the point where the pressure is 0.8 bar.

The condition line may be taken as straight on the h–s diagram.
Answer 2.4 cm, 4.6 m; 3

25 Discuss briefly the reasons for the use of intercooling and aftercooling in reciprocating air compressors.

The following data refers to a two-stage single-acting air compressor:

Air temperatures (°C):	atmospheric	11
	inlet to intercooler	20
	inlet to H.P. cylinder	20
	outlet from aftercooler	20
Air pressures (bar):	atmospheric	1.01
	intercooler	3.28
	delivery	8.0

Compressor shaft speed = 400 rev/min. The diameter of the L.P. and H.P. cylinders are 600 mm and 240 mm respectively and the common stroke is 240 mm. The torque on the compressor shaft is 2.4 kN m for a free air delivery of 0.35 m³/s.

Assuming that the compression and expansion index are equal in both cylinders, estimate the volumetric efficiency, heat removed in the intercooler, and mechanical efficiency.

C_p = 1.005 kJ/kg K.
Answer 0.77; 21.6 kW; 0.80

26 Air is compressed through a pressure ratio of 12:1 in a single-stage reciprocating compressor from an inlet pressure of 1 bar. Calculate the power input for a free air delivery of 4 m³/minute.

Also determine the rotational speed of the crankshaft if the swept volume is 0.02 m³, and the clearance volume is 5 % of the swept volume. The polytropic index for compression and expansion is 1.3.
Answer 22.8 kW; 194 rev/min

27 The compressor in Problem 26 is converted into a two-stage machine with complete intercooling.

Determine the optimum interstage pressure, and the corresponding power input for the same free air delivery.
Answer   3.46 bar; 19.6 kW

28   Air is compressed from 1 bar to 9 bar in a single-stage reciprocating machine. The index of compression and expansion is 1.3 and the clearance volume is 4 % of the swept volume.

Show that the heat transfer during compression is approximately 4.8 times that during expansion.

29   A single-stage air motor uses 0.20 kg/s of compressed air, supplied at 5.5 bar, 120 °C. It exhausts at 1.2 bar. Expansion index = 1.35. Cut-off is at half-stroke.

Neglecting the effects of clearance, determine the mean effective pressure of the cycle and the power developed.
Answer   3.24 bar, 13.3 kW

7

Heat transfer

The transfer of heat is of importance in many fields of energy conversion involving the use of thermal energy: for example, in boilers and furnaces, steam condensers, engines and gas turbines, heat exchangers, waste heat boilers, solar collectors. The science of heat transfer supplements the laws of thermodynamics, enabling the prediction of energy transfer rates. The transfer of heat can be broadly divided into three modes: conduction (solids, fluids), convection (fluids and surfaces), and radiation.

Conduction

For steady-state one-dimensional conduction (i.e. independent of time) the basic law for a plane surface is the Fourier law:

$$Q = KA \frac{dT}{dx}$$

where K = coefficient of thermal conductivity, A = area normal to the direction of the heat flow. In the case of a circular pipe the equation becomes

$$Q = 2\pi K(T_i - T_o)/\ln \frac{r_o}{r_i}$$

per unit length, where the subscripts i and o denote inner and outer.

For a sphere

$$Q = 4\pi K r_o r_i (T_i - T_o)/(r_o - r_i)$$

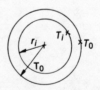

Figure 7.1

If the conduction is **transient** in a plane surface, the Fourier equation becomes

$$\frac{\partial T}{\partial t} = \alpha \frac{\partial^2 T}{\partial x^2}$$

where α = diffusivity = $K/\rho C_p$.

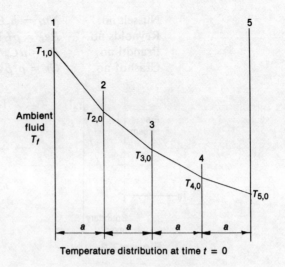

Figure 7.2

Referring to Fig. 7.2 a solution can be obtained by a method of finite differences. For an interior plane

$$T_{i,t} = Fo\,[\,T_{i-1,t-1} + T_{i+1,t-1}\,] + (1 - 2Fo)\,T_{i,t-1}$$

where $T_{i,t}$ = temperature at section i, at time t.

The heat transfer from the ambient fluid to the surface 1 is given by

$$Q = h_c(T_f - T_1)$$

and for that exposed surface the general equation is

$$T_{1,t} = 2Fo(T_{2,t-1} + Bi.T_f) + [1 - 2Fo(1 + Bi)]\,T_{1,t-1}$$

where

Fo = Fourier number = $\alpha \Delta t / a^2$
Δt = time interval
Bi = Biot number = $h_c a / K$.
a = width of layer, as shown in Fig. 7.2.

Convection

Convection is the heat transfer between a fluid and a surface: it may be natural convection, due to buoyancy forces arising from changes in the fluid density, or forced convection due to fluid motion generated by external forces.

The basic law of convection is $Q = h_c A(T_s - T_\infty)$, where h_c (surface coefficient) depends upon the physical properties of the fluid. These properties are measured at the film temperature, $T_f = \frac{1}{2}(T_s + T_\infty)$. The equations used for convection transfer are normally expressed in terms of dimensionless groups (numbers), such as

Nusselt no. $Nu_L = h_c L/K$ based on a characteristic
Reynolds no. $Re_L = \rho L V/\mu$ dimension, L
Prandtl no. $Pr = \mu C_p/K$
Grashof no. $Gr = \rho^2 \beta g D^3 (T_s - T_\infty)/\mu^2$

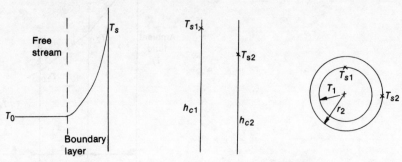

Figure 7.3   **Figure 7.4**

Overall coefficient, U   This takes into account the individual coefficients and overall temperature drop. Referring to Fig. 7.4, the overall coefficient is given by

$$\frac{1}{U} = \frac{1}{h_{c1}} + \frac{x}{K} + \frac{1}{h_{c2}}$$

(plane surface), or

$$\frac{1}{r_1 U_1} = \frac{1}{r_2 U_2} = \frac{1}{r_2 h_{c2}} + \frac{1}{r_1 h_{c1}} + \frac{1}{K} \ln \frac{r_2}{r_1}$$

per unit length (cylindrical surface).

Fins (extended surfaces)   In a fin of constant cross-sectional area (Fig. 7.5) the temperature along the fin is given by the equation

$$\theta = C_1 \exp(-\xi\sqrt{Bi}) + C_2 \exp(\xi\sqrt{Bi})$$

where, $\theta = \dfrac{T_x - T_f}{T_b - T_f}$, $\xi = \dfrac{x}{L}$, $Bi = \dfrac{h_c P L^2}{kA}$, P = perimeter, A = area.

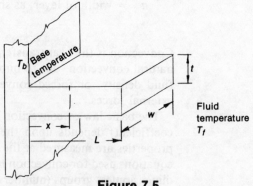

Figure 7.5

126 SOLVING PROBLEMS IN APPLIED THERMODYNAMICS AND ENERGY CONVERSION

Thermal radiation

The performance of a fin is measured by means of the fin efficiency η_f, defined as the actual heat-transfer rate/ideal heat-transfer rate.

$$\eta_f = \text{actual transfer rate}/h_c A_s (T_b - T_f)$$

where A_s = total surface area exposed to the fluid.

Heat transfer by radiation is part of the electromagnetic spectrum, and does not require the presence of a solid or fluid material. The basic laws and terms used are summarized as follows:

Stefan–Boltzmann law: $E_b = \sigma A T^4$.

Kirchhoff's law: the absorptivity of a body, α = the emissivity, e at thermal equilibrium. The monchromatic values $\alpha_\lambda = e_\lambda$ are not restricted to thermal equilibrium.

Emissivity e = energy emitted by the body/energy emitted by a black body at the same temperature.

Geometrical (shape) factor F_{ij} = fraction of radiant energy emitted from black surface i which arrives directly on black surface j.

A grey surface is one in which the monochromatic and total properties are the same, $e_\lambda = e$, $\alpha_\lambda = \alpha$.

For a grey gas the reflectivity is zero. The transmittivity $\tau_g = 1 - \alpha_g = 1 - e_g$. The emissivity of a gas depends upon the partial pressure, temperature and beam length $L = 3.6 \times$ volume/surface area.

Electrical analogy

In the electrical analogy temperature = potential, thermal resistance = electrical resistance, and heat flow = electrical current.

The analogies are shown in Fig. 7.6, where x = thickness, K = thermal conductivity, A = area, h_c = convection coefficient, F = geometrical factor.

In the case of a grey gas filling the space between two grey surfaces the electrical analogy is shown in Fig. 7.7.

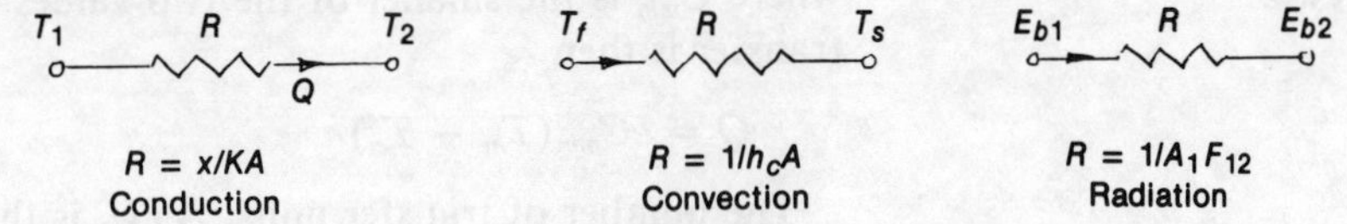

$T_1 \xrightarrow{R} T_2 \qquad T_f \xrightarrow{R} T_s \qquad E_{b1} \xrightarrow{R} E_{b2}$
Q

$R = x/KA$ $\qquad R = 1/h_c A$ $\qquad R = 1/A_1 F_{12}$
Conduction $\qquad$ Convection $\qquad$ Radiation

Figure 7.6

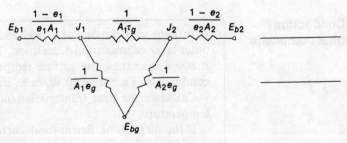

Figure 7.7

Heat exchangers

The exchange of heat in an exchanger involves a temperature difference between the fluids, which varies along the path. A single-pass contraflow exchanger is shown in Fig. 7.8.

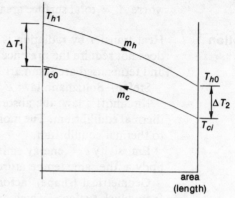

Figure 7.8

The heat-transfer rate over the whole area $= \int U \, dA \, dT$, and if the overall coefficient is constant, this becomes $UA \, \Delta T_m$. The mean temperature difference is given by $\Delta T_m = (\Delta T_1 - \Delta T_2)/\ln(\Delta T_1/\Delta T_2)$.

For more complex exchangers, with several passes, or cross-flow paths, the log mean temperature difference is multiplied by a correction factor.

An energy balance on the exchanger gives

$$m_h C_{ph}(T_{hi} - T_{ho}) = m_c C_{pc}(T_{co} - T_{ci})$$

where m = mass flow rate
C_p = specific heat

Effectiveness and NTU

The effectiveness $e = C_{ph}(T_{hi} - T_{ho})$ or $C_{pc}(T_{co} - T_{ci})/C_{min}(T_{hi} - T_{ci})$ where C_{min} is the smaller of the two values $m_h C_{ph}$ and $m_c C_{pc}$. The heat transfer is then

$$Q = eC_{min}(T_{hi} - T_{ci})$$

The number of transfer units, *NTU*, is the ratio UA/C_{min}. Graphs of effectiveness plotted against *NTU* for different values of C_{min}/C_{max} are available for a wide variety of heat-exchanger geometries.

7.1 Conduction through a composite wall

A furnace wall consists of a steel plate, 1 cm thick, covered by an outer layer of insulation 5 cm thick. The inner surface temperature is 800 K, and the outer surface temperature is 350 K. The thermal conductivities are steel 20 W/m K, insulation 1 W/m K.

Calculate the heat transmission/m^2 surface area, and interface temperature.

If the surface coefficient (both surfaces) is 50 W/m^2 K, calculate the hot gas (inner) and air temperatures.

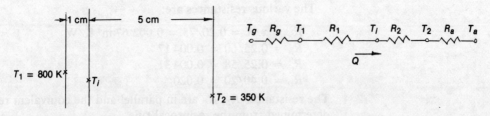

Figure 7.9

Solution Referring to Fig. 7.9, per unit area, the resistances are

$$R_g = \frac{1}{50}, \quad R_1 = \frac{x}{K} = \frac{0.01}{20}, \quad R_2 = \frac{0.05}{1}, \quad R_a = \frac{1}{50}$$

Therefore $Q = \dfrac{T_g - T_a}{R_g + R_1 + R_2 + R_c} = \dfrac{T_1 - T_2}{R_1 - R_2} = \dfrac{800 - 350}{\dfrac{0.01}{20} + 0.05}$

$= 8911 \text{ W/m}^2$.

Also $Q = \dfrac{T_g - T_1}{R_g}$, therefore, $T_g = 8911\left(\dfrac{1}{50}\right) = $ **978 K**

$Q = (T_2 - T_a)/R_a$, therefore $T_a = 350 - \dfrac{8911}{50} = $ **172 K**.

7.2 Composite wall

A composite wall is shown in Fig. 7.10. The surface temperature of wall A is 500 °C, and of D is 100 °C. Assuming one-dimensional flow calculate the heat transmission/m² and interface temperatures.

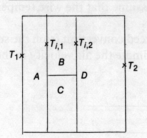

Figure 7.10

Figure 7.11

Material	Conductivity K (W/m K)	Thickness x (cm)
A	75	20
B	60	25
C	58	25
D	20	40

Solution The electrical analogy is shown in Fig. 7.11. For one-dimensional flow the interface temperatures between A and B and between A and C will be equal. Those between B and D and between C and D will also be equal.

HEAT TRANSFER 129

The various resistances are:

$R_A = x/K = 0.20/75 = 0.00267 \text{ m}^2 \text{ K/W}$
$R_B = 0.25/60 = 0.00417$
$R_C = 0.25/58 = 0.00431$
$R_D = 0.40/20 = 0.020$

The resistances R_B, R_C are in parallel and the equivalent resistance R_e is determined from the reciprocal rule

$$\frac{1}{R_e} = \frac{1}{R_B} + \frac{1}{R_C} = 472, \text{ therefore } R_e = 0.00212$$

Hence the heat transmission $Q = \dfrac{T_1 - T_2}{R_A + R_e + R_D} = \dfrac{500 - 100}{0.02479}$

$= \mathbf{16\,136 \text{ W/m}^2}$

The interface temperatures can now be calculated:

$$Q = \frac{T_1 - T_{i1}}{R_A}$$

Therefore, $T_{i1} = 500 - 0.00267(16\,136) = \mathbf{457\,°C}$, and
$Q = (T_{i2} - T_2)/R_D$
Therefore, $T_{i2} = 100 + 0.020(16\,136) = \mathbf{423\,°C}$.

7.3 Conduction through a circular section

A 1 mm diameter wire is covered with a 2 mm thickness layer of insulation ($K = 0.5$ W/m K). The ambient temperature of the air is 25 °C and the surface conductance is 10 W/m² K. The temperature of the wire is 100 °C.

Calculate the heat loss from the wire/m length with and without the insulation. Assume that the wire temperature is not affected by the insulation.

The loss by forced convection from the surface is given by $Nu_D = C.Re^n.Pr^{0.33}$. Estimate the air velocity in the insulated case.

Table 7.1

Re_D	C	n
0.4–4	0.99	0.33
4–40	0.91	0.38
40–4000	0.68	0.47

Solution The Biot number $= h_c r_o/K = 10(2 + 0.5) \times 10^{-3}/0.5 = 0.05$; therefore, since it is less than 1, the presence of insulation will *increase* the heat loss.

With insulation $Q = \dfrac{100 - 25}{\dfrac{\ln(2.5/0.5)}{2\pi(0.5)} + \dfrac{1}{2\pi(0.0025)(10)}} = \mathbf{10.90 \text{ W/m}}$

Without insulation $Q = h_c A(T - T_\infty)$
$$= 10 \times 2\pi(0.005)(100 - 25) = \textbf{2.36 W/m}$$

With the insulation: heat transfer through the insulation = heat loss from the surface (at a temperature T_s).

Therefore, $10.90 = (T_s - 25) \times 2\pi(0.0025)(10)$, therefore $T_s = 94$ °C.

The air properties are taken from tables at the film temperature $T_f = \frac{1}{2}(T_s + T_\infty) = (94 + 25)/2 = 60$ °C. The values are

$K = 0.029$ W/m^2 K
$Pr = 0.70$
$v = 1.93 \times 10^{-5}$ m^2/s

Therefore, $Re_D = \dfrac{vD}{v} = \dfrac{v \times 0.0025}{1.93 \times 10^{-5}} = 129.5\,v$

$Nu_D = \dfrac{h_c D}{K} = \dfrac{10 \times 0.0025}{0.029} = 0.86 = c(129.5\,v)^n (0.70)^{0.33}$

Therefore, $c(129.5\,v)^n = 0.97$.

Since c and n depend upon the Reynolds number, Re, a range of Re must be assumed and then checked. Assuming $4 < Re < 40$, $0.91(129.5\,v)^{0.38} = 0.97$ giving $v = 0.009$ m/s. This gives $Re_D = 1.16$, which is not in the range assumed.

Now assume that $0.4 < Re_D < 4$, therefore $0.99(129.5\,v)^{0.33} = 0.97$, giving $v = 0.0076$ m/s. This gives $Re_D = 129.5 \times 0.0076 = 0.98$, which falls in the range assumed. Hence $v = \textbf{7.6 mm/s}$.

7.4 Finned surface

A steel fin (conductivity 20 W/m K) has a circular cross-section of diameter 2 cm, length 10 cm; and is attached to a wall which is at a temperature of 300 °C. The end of the fin is insulated. The temperature of the fluid in contact with the fin is 50 °C, and the convection coefficient is 10 W/m^2 K.

Calculate the heat dissipated from the fin, the temperature at the end of the fin, and the fin efficiency.

Solution The fin is shown in Fig. 7.12. The area

$$A = \frac{\pi}{4}(0.02)^2 = 3.142 \times 10^{-4} \text{ m}^2$$

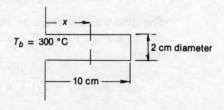

Figure 7.12

and perimeter $P = \pi(0.02)$.
Therefore, Biot number $Bi =$

$$\frac{h_c P L^2}{kA} = \frac{10 \times 0.02\pi \times (0.1)^2}{20\pi \times 10^{-4}} = 1.0$$

The general solution for the temperature distribution is

$$\theta = C_1 \exp(-\xi\sqrt{Bi}) + C_2 \exp(\xi\sqrt{Bi})$$

where, $\theta = \dfrac{T_x - T_f}{T_b - T_f}$, $\xi = \dfrac{x}{L}$

In this problem the boundary conditions are

(a) $dT/dx = 0$ at $x = L$ (the fin is insulated at the tip) or $d\theta/d\xi = 0$ at $\xi = 1$.
(b) $T = T_b$ at $x = 0$, or $\theta = 1$ at $\xi = 0$ (the fin temperature at the base is equal to the wall temperature).

Hence, substituting these boundary conditions into the equations
$\theta = C_1 \exp(-\xi\sqrt{Bi}) + C_2 \exp(\xi\sqrt{Bi})$
$d\theta/d\xi = -C_1\sqrt{Bi} \exp(-\xi\sqrt{Bi}) + C_2\sqrt{Bi} \exp(\xi\sqrt{Bi})$
gives $0 = -C_1\sqrt{Bi} \exp(-\sqrt{Bi}) + C_2\sqrt{Bi} \exp(\sqrt{Bi})$
$1 = C_1 \exp(0) + C_2 \exp(0) = C_1 + C_2$.

Solving for the constants

$$C_1 = [1 + \exp(-2\sqrt{Bi})]^{-1}$$

$$C_2 = \frac{\exp(-2\sqrt{Bi})}{1 + \exp(-2\sqrt{Bi})}$$

and substituting

$$\theta = \frac{\exp(-\xi\sqrt{Bi})}{1 + \exp(-2\sqrt{Bi})} + \frac{\exp(-2\sqrt{Bi})\exp(\xi\sqrt{Bi})}{1 + \exp(-2\sqrt{Bi})}$$

$$= \frac{\exp(-\xi\sqrt{Bi}) + \exp(-2\sqrt{Bi})\exp(\xi\sqrt{Bi})}{1 + \exp(-2\sqrt{Bi})}$$

$$= \frac{\exp(\sqrt{Bi})\exp(-\xi\sqrt{Bi}) + \exp(-\sqrt{Bi})\exp(\xi\sqrt{Bi})}{\exp(\sqrt{Bi}) + \exp(-\sqrt{Bi})}$$

$$= \cosh[(1-\xi)\sqrt{Bi}]/\cosh\sqrt{Bi}$$

The heat transfer from an element dx, at a distance x from the base is
$dQ = h_c P \, dx \, (T_x - T_f) = h_c P \, dx \, \theta(T_b - T_f)$, therefore over the fin length

$$Q = h_c P(T_b - T_f) \int_0^L \theta \, dx$$

$$= \frac{h_c P(T_b - T_f)L}{\cosh\sqrt{Bi}} \int_0^1 \cosh[(1-\xi)\sqrt{Bi}] \, d\xi$$

(since $\xi = x/L$, then $dx = L\, d\xi$). Therefore

$$Q = \frac{h_c PL(T_b - T_f)}{\cosh \sqrt{Bi}} \left[-\frac{1}{\sqrt{Bi}} \sinh\left[(1 - \xi)\sqrt{Bi}\right] \right]_0^1$$

$$= \frac{h_c PL}{\sqrt{Bi}}(T_b - T_f) \tanh \sqrt{Bi}$$

$$= \frac{kA}{L}\sqrt{Bi}(T_b - T_f) \tanh \sqrt{Bi}$$

since $\dfrac{h_c PL}{\sqrt{Bi}} = \dfrac{h_c PL \sqrt{Bi}}{\sqrt{Bi}} = h_c PL\sqrt{Bi}\,\dfrac{kA}{h_c PL^2} = \dfrac{kA}{L}\sqrt{Bi}$

In this problem $Bi = 1$, therefore

$$Q = \frac{20(\pi \times 10^{-4})}{0.1} \times (300 - 50) \tanh 1$$

$$= 15.708 \times 0.7616 = \mathbf{11.96\ W}.$$

At the end of the fin, $\xi = 1$, therefore $\theta = \cosh 0/\cosh 1 = 0.648$ and $T_x = T_f + \theta(T_b - T_f) = 50 + 0.648(300 - 50) = \mathbf{212\ °C}$.

The ideal heat transfer rate corresponds to a fin temperature T_b along the whole length of the fin, and is therefore $h_c PL(T_b - T_f) = 10\pi \times 0.02 \times 0.1 \times 250 = 15.71$ W.

Therefore fin efficiency $= 11.96/15.71 = \mathbf{0.76}$.

7.5 Fin efficiency

Briefly outline the purpose of using a finned tube instead of a plain tube in a heat exchanger. Define the term fin efficiency.

A tube is fitted with an annular fin, thermal conductivity 200 W/m K. The outside diameter of the tube is 8 cm; the fin is of rectangular cross-section, 5 mm thick, 16 cm outer diameter. Convection coefficient = 60 W/m² K. Base temperature of the fin = 250 °C, and the surrounding fluid temperature is 70 °C.

Determine the heat transfer rate from the fin. The fin tip is not insulated.

A fin efficiency curve, based on the corrected length L_c, is provided.

Solution The fin efficiency curve is shown in Fig. 7.13.

The addition of a fin (or extended surface) to the plain surface of a tube considerably increases the heat exchange area between the surface and fluid, much more effectively than simply increasing the diameter. In the annular fin shown, the heat exchange area is increased from $2\pi r_1 t$ to $\pi(r_2^2 - r_1^2) + 2\pi r_2 t$.

In an ideal fin the material temperature would be equal to the base temperature and the heat dissipated from the fin would be $h_c A(T_b - T_f)$. In a real fin the material conductivity is finite and there is a temperature

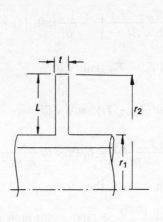
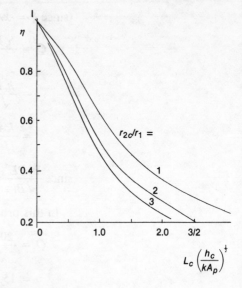

Figure 7.13

drop along the fin length. The heat exchange rate is therefore less than the ideal, and the ratio of the two rates is the fin efficiency.

The fin tip is not insulated in this problem, so that a corrected length is used. $L_c = L + \frac{1}{2}t = 4 + \frac{1}{2}(\frac{1}{2}) = 4.25$ cm. The profile area $A_p = L_c t = 2.125$ cm². Therefore

$$L_c^{3/2}\left(\frac{h_c}{KA_p}\right)^{1/2} = \left(\frac{4.25}{100}\right)^{3/2}\left(\frac{60}{200 \times 2.125 \times 10^{-4}}\right)^{1/2} = 0.33$$

and $r_{2c}/r_1 = (r_1 + L_c)/r_1 = (4 + 4.25)/4 = 1.063$; therefore from the graph, fin efficiency = 0.89, so heat-transfer rate

$$Q = \eta h_c A(T_b - T_f)$$
$$= 0.89 \times 60 \times \pi[(0.0825)^2 - (0.04)^2](250 - 70)$$
$$= \mathbf{314\ W}$$

The heat-transfer rate from the plain tube is $60 \times 2\pi(0.04)(0.005)(250 - 70) = 13.6$ W, so that the effect of the fin can be seen: the rate is increased by a factor of 23.

7.6 Transient conduction

A large steel plate, initially at a uniform temperature of 300 °C, is cooled by blowing air at 50 °C over the plate surface. The plate is 10 cm thick; its thermal conductivity is 40 W/m K, and diffusivity $\alpha = 10^{-5}$ m²/s. The convection coefficient is 400 W/m² K.

Determine the time taken for the surface temperature to decrease to 200 °C, and the temperature at planes 1 cm and 10 cm from the surface at that time.

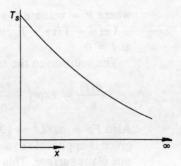

Figure 7.14

Solution In the case of transient conduction the Fourier equation becomes

$$\frac{\partial^2 T}{\partial x^2} = \frac{1}{\alpha}\frac{\partial T}{\partial x}$$

The initial and boundary conditions are

$T(x, 0) = T_0$

$T(\infty, t) = T_0$

$h_c[T_\infty - T(0, t)] = -K\left(\frac{\partial T}{\partial x}\right)_{x=0}$

The analytical solution is

$$\frac{T(x, t) - T_0}{T_\infty - T_0} = 1 - \text{erf } \xi$$

$$- [\exp(Bi + \eta)][1 - \text{erf}(\xi + \sqrt{\eta})]$$

where $\text{erf }\sqrt{\xi} = \frac{2}{\sqrt{\pi}}\int_0^{\sqrt{\xi}} \exp(-\eta^2)\, d\eta$

and $\xi = \frac{x^2}{4\alpha t} = \frac{1}{2}Fo^{-1/2}$

$Fo = \alpha t/x^2$

$Bi = h_c x/K$

$\eta = h_c^2 \alpha t/K^2 = (Bi)^2 Fo$

In this problem $Bi = h_c L/K = 400 \times 0.10/40 = 1.0$. The Biot number is the ratio of the conductive to convective resistance, and if $Bi \ll 1.0$ the transient temperature can be obtained from an energy balance.

The decrease in stored energy in the solid = heat transfer rate from the surface by convection. Therefore

$$-\rho V C_p \frac{dT}{dt} = h_c A_s[T(t) - T_\infty]$$

where V = volume of solid, A_s = surface area.

Let $\theta = T(t) - T_\infty$, and initially $T = 0$ at time $t = 0$, i.e. $\theta_0 = T_0 - T_\infty$ at $t = 0$.

The solution to the equation is then

$$\frac{\theta(t)}{\theta_0} = \exp\left(-\frac{h_c A_s}{\rho V C_p}\right) t = \exp - (Bi)(Fo)$$

Also $Fo = \alpha t/L^2 = 10^{-5} t/(0.1)^2 = 10^{-3} t$. The solution of the equation given is applicable to a semi-infinite geometry, namely a large body with one plane surface. This condition requires that $Fo < 1.0$. Hence the plate can be considered as semi-infinite provided $t < 10^3$ s.

The first part of the problem requires the determination of the time taken for the surface temperature (at $x = 0$) to decrease to 200 °C, that is when

$$\frac{T(0, t) - T_0}{T_\infty - T_0} = \frac{200 - 300}{50 - 300} = 0.40$$

At $x = 0$, $\xi = 0$, therefore

$$0.40 = 1 - \text{erf } 0 - [\exp(1 + \eta)][1 - \text{erf } \sqrt{\eta}].$$

From tables of the error function, $\text{erf } 0 = 0$, therefore

$$[\exp(1 + \eta)][1 - \text{erf }\sqrt{\eta}] = 0.60.$$

Solving gives $\eta = 0.25$, therefore $Bi\sqrt{Fo} = \sqrt{\eta} = 0.5$, therefore $t = \eta K^2/\alpha h_c^2 = 0.25(40)^2/10^{-5}(400)^2 =$ **250 s**.

The second part of the problem requires the temperature, at 250 s, at the planes $x = 1$ cm and $x = 10$ cm.

x (cm)	ξ	$\sqrt{\eta}$	$\theta(x, t)/\theta_\infty$	$T(x, t)$
1	0.01	0.5	0.325	**219 °C**
10	1.0	0.5	0.039	**290 °C**

7.7 Transient conduction: numerical solution

A large, thick plate is at a uniform temperature of 200 °C. The plate surface is suddenly exposed to a fluid at 100 °C. The convection coefficient between the fluid and plate surface is 500 W/m² K. For the plate $K = 40$ W/m K, $\alpha = 3 \times 10^{-5}$ m²/s.

Determine the temperature distribution at the surface and at 4, 8, 12, 16 cm from the surface for the first minute after exposure. A numerical method should be used.

Solution The requisite equations are

$$T_{1,t+\Delta t} = 2(Fo)[T_{2,t} + (Bi)T_\infty] + [1 - 2(Fo) - 2(Fo)(Bi)] T_{1,t}$$

and for the interior nodes

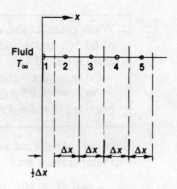

Figure 7.15

$$T_{i,t+\Delta t} = (Fo)(T_{i-1,t} + T_{i+1,t}) + [1 - 2(Fo)] T_{i,t}$$

For a stable solution, the limits are

surface node $Fo(1 + Bi) \leq \frac{1}{2}$
internal nodes $Fo \leq \frac{1}{2}$

In this problem Δx is given as 4 cm and

$$Fo = \alpha \Delta t/(\Delta x)^2 = 3 \times 10^{-5} \Delta t/(0.04)^2 = 0.01875 \Delta t,$$

and $Bi = h_c \Delta x/K = \frac{1}{2}$.
Therefore $Fo \leq 0.5/1.5 \leq \frac{1}{3}$.
Taking $Fo = \frac{1}{4}$ gives a suitable $\Delta t = 13.33$ s.

The surface node temperature is then determined from

$$T_{1,t+\Delta t} = \frac{1}{2}(T_{2,t} + \frac{1}{2}T_\infty) + \frac{1}{4} T_{1,t}$$

and the interior node temperatures from

$$T_{i,t+\Delta t} = \frac{1}{4}(T_{i-1,t} + T_{i+1,t}) + \frac{1}{2} T_{i,t}$$

Table 7.2 shows the values obtained.

Table 7.2

Time	Node 1	2	Temperature (°C) 3	4	5
(s)	$x = 0$	4	8	12	16 cm
0	200	200	200	200	200
13.33	175	200	200	200	200
26.67	168.8	193.9	200	200	200
40.0	164.1	189.1	198.5	200	200
53.33	160.6	185.2	196.5	199.6	200
66.67	157.8	181.9	194.5	198.9	199.9

7.8 Forced convection

> Water enters a tube, diameter 10 mm, length 50 cm, at a temperature of 60 °C, and a velocity of 10 cm/s. The tube wall temperature is constant at 80 °C.
>
> Estimate the water temperature at the tube exit.
>
> $Nu_D = 1.86(Re_D Pr)^{0.33}(D/L)^{0.33}(\mu_b/\mu_s)^{0.14}$ for laminar flow, provided that $Re_D Pr D/L > 10$.

Solution At the tube inlet, $T_w = 333$ K and the properties of water at this temperature are

$$v = 0.48 \times 10^{-6} \text{ m}^2/\text{s}, Pr = 3.0$$

therefore, $Re_D = vD/v = 0.10(0.01)/(0.48 \times 10^{-6}) = 2083$, therefore the flow is laminar.

Also $Re_D Pr D/L = 2083 \times 3 \times 0.01/0.5 = 125$ and is greater than 10; hence the equation given is applicable.

The equation is based upon the properties of the water at the mean film temperature, $T_f = \frac{1}{2}(T_w + T_s)$, where T_w = mean water temperature = $\frac{1}{2}(T_{w1} + T_{w2})$.

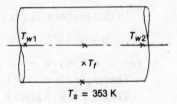

Figure 7.16

The heat transfer is

$$Q = mC_p(T_{w2} - T_{w1})$$
$$= h_c A(T_s - T_w)$$

Since the water temperature at the exit is unknown, the mean bulk temperature T_w is unknown and the water properties cannot be determined. An iterative solution is therefore necessary.

Try $T_{w2} = 350$ K.

Therefore, $T_w = \frac{1}{2}(333 + 350) = 342$ K.

From tables, $K = 0.665$ W/m K, $Pr = 2.63$, $\mu = 412 \times 10^{-6}$ Ns/m², $C_p = 4187$ J/kg K, $\rho = 978$ kg/m³, and substituting $Re_D = 2374$,

$$Nu_D = 1.86(2374 \times 2.63)^{0.33}(0.01/0.5)^{0.33}$$
$$\times (412 \times 10^{-6}/352 \times 10^{-6})^{0.14}$$

where μ_b = viscosity of water at the mean bulk temperature
μ_s = viscosity of water at the surface temperature.

Therefore, $Nu_D = 9.626$ and $h_c = K Nu_D/D = 0.665(9.626)/0.01 = 640$ W/m² K.

The mass flow rate $m = \rho v(\pi D^2)/4 = 0.00768$ kg/s; therefore

$$Q = 640 \times \pi(0.01)(0.5)(353 - 342)$$
$$= 0.00768 \times 4187(T_{w2} - 333)$$

giving $T_{w2} = 336$ K (compared to the assumed value 350 K).
A second iteration taking $T_{w2} = 341$ K gives

$$T_w = 337 \text{ K}$$
$$K = 0.657 \text{ W/m K}, Pr = 2.78, \mu = 436 \times 10^{-6} \text{ N s/m}^2$$
$$C_p = 4188 \text{ J/kg K}, \rho = 980 \text{ kg/m}^3$$

and
$$Re_D = 2243$$
$$Nu_D = 1.86 \times 17.878 \times 0.2750 \times 1.0304 = 9.423$$
$$h_c = 619 \text{ W/m}^2 \text{ K}$$

therefore
$$Q = 619\pi \times 0.01 \times 0.5(353 - 337)$$
$$= 0.00768 \times 4188(T_{w2} - 333)$$

giving $T_{w2} = \mathbf{338}$ **K** (close to assumed value of 341 K).

7.9 Free convection

> The variables involved in natural (free) convection from a surface at temperature T_s to a fluid at temperature T_∞ can be taken as the fluid properties: density ρ, viscosity μ, coefficient of thermal expansion β, thermal conductivity K, specific heat C_p; a characteristic dimension of the surface D, and the temperature difference $\Delta T = T_s - T_\infty$.
>
> Show, by dimensional analysis, an equation of the form $Nu = \phi(Gr, Pr)$ is applicable.
>
> A pipe passes horizontally through a room containing still air at 20 °C. The pipe outer diameter is 70 mm, and its surface temperature 100 °C. The exposed length of pipe is 3 m. Calculate the heat loss by convection from the pipe. $\beta = 0.0037$. $Nu_D = 0.59(Gr_D Pr)^{0.25}$.

Solution The groups quoted are the Nusselt, Grashof and Prandtl numbers.
Assume that

$$h_c = \rho^a \mu^b \beta^c K^d C_p^e D^f \Delta T^h g^i$$

The method of dimensional analysis uses the fact that if the equation is to be physically valid, the dimensions of each side must be the same. An equation, for example 1 m = 6 kg, is in this form meaningless.

Hence if all the variables are expressed in terms of primary dimensions mass M, length L, time t, temperature θ and heat Q, the requirement of the same dimensions on each side of the equation gives suitable groupings. It should be noted that five dimensions are used for convenience rather than the three fundamental dimensions of mass, length and time.

The variables and their dimensions are listed:

density	ρ	ML^{-3}
viscosity	μ	$ML^{-1}t^{-1}$
expansion	β	θ^{-1}

conductivity	K	$QL^{-1}\theta^{-1}t^{-1}$
specific heat	C_p	$QM^{-1}\theta^{-1}$
dimension	D	L
temp. diff.	ΔT	θ
acceleration	g	Lt^{-2}
conv. coefficient	h_c	$QL^{-2}\theta^{-1}t^{-1}$

Substituting:

$$QL^{-2}\theta^{-1}t^{-1} = (ML^{-3})^a(ML^{-1}t^{-1})^b(\theta^{-1})^c \\ \times (QL^{-1}\theta^{-1}t^{-1})^d(QM^{-1}\theta^{-1})^e(L)^f \\ \times (\theta)^h(Lt^{-2})^i$$

Equating the powers of each dimension:

$$\begin{aligned} M & - & 0 &= a + b - e \\ L & - & -2 &= -3a - b - d + f + i \\ t & - & -1 &= -b - d - 2i \\ \theta & - & -1 &= -c - d - e + h \\ Q & - & 1 &= d + e \end{aligned}$$

Expressing a, b, c, d, f, h in terms of e, i

$$\begin{aligned} a &= 2i & d &= 1 - e & h &= i \\ b &= e - 2i & f &= 3i - 1 & c &= i \end{aligned}$$

Substituting

$$h_c = \rho^{2i}\mu^{e-2i}\beta^i K^{1-e} C_p^e D^{3i-1}\Delta T^i g^i$$

$$= \frac{K}{D}\left(\frac{\mu C_p}{K}\right)^e \left(\frac{\rho^2 \beta D^3 g \, \Delta T}{\mu^2}\right)^i$$

or $\dfrac{h_c D}{K} = Nu_D = \phi(Pr, Gr_D)$

This equation shows that the Nusselt number is a function of the Grashof number Gr and Prandtl number Pr. The former number is the ratio of the buoyancy force $\rho g \beta(T_s - T_\infty)$ to the shear force, and the latter is the ratio of the kinematic viscosity to the diffusivity.

The fluid temperature $T_\infty = 20\,°C$ and the surface temperature $T_s = 100\,°C$, and the air properties are evaluated at the film temperature $T_f = \frac{1}{2}(100 + 20) = 60\,°C$.

$$\begin{aligned} C_p &= 1007\text{ J/kg K} & K &= 0.029\text{ W/m K} \\ \mu &= 2.02 \times 10^{-5}\text{ N s/m}^2 & Pr &= 0.70 \\ \rho &= 1.047\text{ kg/m}^3 & & \\ Gr_D &= \rho^2 g\beta D^3(T_s - T_\infty)/\mu^2 = 4.455 \times 10^7 & & \\ Gr_D Pr &= 3.12 \times 10^7 & & \end{aligned}$$

Hence $Nu_D = 0.59(3.12 \times 10^7)^{0.25} = 44.1 = h_c(0.07)/0.029$, therefore $h_c = 18.27\text{ W/m}^2\text{ K}$ and the heat loss
$= h_c \pi DL(T_s - T_\infty)$
$= 18.27 \times \pi \times 0.07 \times 3 \times 80 =$ **964 W**.

7.10 Solar radiation

A black plane is buried so that its upper surface is level with the earth's surface, and the lower surface is perfectly insulated. The air temperature is 27 °C. The convection coefficient = 10 W/m² K.

Calculate the equilibrium temperature of the plane at the two conditions:

(a) clear night sky, equivalent to a black body at 100 K,
(b) cloudy night sky, equivalent to a black body at 250 K.

Comment on the values obtained.

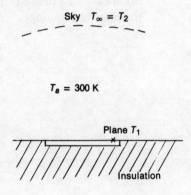

Figure 7.17

Solution The net heat flow into the plane from the air is $Q_i = h_c A_1 (T_a - T_1)$, and the net flow from the plane to the sky is $Q_o = \sigma A_1 (T_1^4 - T_2^4)$. For thermal equilibrium $Q_i = Q_o$. Therefore $h_c(T_a - T_1) = \sigma(T_1^4 - T_2^4)$.

In case (a): $10(300 - T_1) = 5.67 \times 10^{-8}(T_1^4 - 100^4)$ giving $T_1 = 270$ K = -3 °C.

In case (b): $10(300 - T_1) = 5.67 \times 10^{-8}(T_1^4 - 250^4)$ giving $T_1 = 285$ K = 12 °C.

The values obtained show that the plane can attain frost conditions even though the air temperature is well above 0 °C, due to the good view of a very clear sky. This effect is well known to farmers and gardeners!

The presence of cloud increases the effective sky temperature since clouds absorb heat and re-radiate to the earth's surface. The main absorbers of radiation, in the pertinent wavelengths of the spectrum, are CO_2 and water vapour. Hence the so-called 'greenhouse effect' due to the emission of CO_2 from combustion of fuels.

The effective sky temperature is also increased by the presence of surrounding objects such as buildings, trees, etc. Thus a car windscreen may be covered by frost on a clear night when parked on open ground, but not if shielded by a building (or tree).

7.11 Radiation, two surfaces

Two grey opaque surfaces form an enclosure. Show that the heat input to surface 1 is given by

$$Q_1 = \sigma(T_1^4 - T_2^4) / \left[\frac{1 - e_1}{e_1 A_1} + \frac{1}{A_1 F_{12}} + \frac{1 - e_2}{e_2 A_2} \right]$$

The surfaces are two long circular pipes. The inner pipe is 80 cm diameter (emissivity $e = 0.5$), and the outer pipe is 1 m diameter ($e = 0.3$). The surface temperature of the inner pipe is 550 K, and of the outer pipe 300 K.

Calculate the net heat transfer/m length of pipe.

Solution The desired equation can be obtained from a consideration of the electrical analogy.

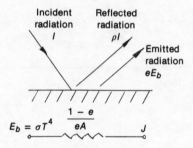

Figure 7.18

For a grey opaque surface (Fig. 7.18) $\alpha = e = 1 - \rho$, and radiosity $J = eE_b + \rho I = eE_b + (1 - e)I$. The net flow to the surface is

$$Q = A(J - I)$$
$$= A[J - (J - eE_b)/(1 - e)]$$
$$= e(E_b - J)/(1 - e)$$
$$= (E_b - J)/R$$

where $R = (1 - e)/eA$.

The electrical analogy for two surfaces is shown in Fig. 7.19.

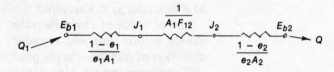

Figure 7.19

For thermal equilibrium the heat supplied to surface 1 from outside the system, Q_1 = net heat transfer between the surfaces, Q_{12} = the heat flow from surface 2, Q_2, i.e.

$$Q_1 = Q_{12} = Q_2$$
$$= (E_{b1} - E_{b2})/R$$

where R = total resistance = $\dfrac{1-e_1}{e_1 A_1} + \dfrac{1}{A_1 F_{12}} + \dfrac{1-e_2}{e_2 A_2}$

$$E_b = \sigma T^4$$

Hence the desired equation follows.

In this problem, the geometrical (shape) factor from surface 1 to surface 2 is 1 since the cylinders are long, and end effects are neglected.

$$E_{b1} = 5.67 \times 10^{-8}(550)^4 = 5187 \text{ W/m}^2, \quad e_1 = 0.5$$
$$E_{b2} = 5.67 \times 10^{-8}(300)^4 = 459 \text{ W/m}^2, \quad e_2 = 0.3$$

therefore, $Q_{12} = (5187 - 459)/\left[\dfrac{0.5}{0.5 \times 0.8\pi} + \dfrac{1}{0.8\pi} + \dfrac{0.7}{0.3\pi}\right]$

= **3073 W/m** length.

7.12 Radiation between black surfaces

Two directly opposed rectangles, 2 m × 4 m, are spaced 1 m apart. The surfaces are black, and maintained at 500 K and 400 K. The backs of both surfaces are insulated, and there is no incident energy from other sources. The geometrical factor between the surfaces = 0.52.

Determine a heat balance for the system.

Solution The electrical analogy for three surfaces is shown in Fig. 7.20.

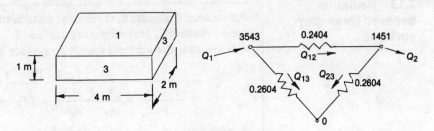

Figure 7.20

In this problem the incident radiation from the surroundings is zero, so that the surroundings can be considered as a black body at 0 K.

$F_{12} = 0.52$ therefore $F_{13} = 1 - 0.52 = 0.48$
$A_1 F_{12} = A_2 F_{21}$ therefore $F_{21} = F_{12} = 0.52$
$A_1 F_{13} = A_2 F_{23}$ (symmetry) therefore $F_{23} = 0.48$
$E_{b1} = \sigma T_1^4 = 5.67 \times 10^{-8}(500)^4 = 3543 \text{ W/m}^2$
$E_{b2} = 5.67 \times 10^{-8}(400)^4 = 1451 \text{ W/m}^2$
$E_{b3} = 0$

$1/A_1F_{12} = 1/(8 \times 0.52) = 0.2404$
$1/A_1F_{13} = 1/(8 \times 0.48) = 0.2604$
Q_{12} = net heat transfer between surfaces 1 and 2
$= (3543 - 1451)/0.2404 = 8702$ W
Q_{13} = net heat transfer between surfaces 1 and 3
$= 3543/0.2604 = 13606$ W
$Q_{23} = 1451/0.2604 = 5572$ W.

The heat supplied to each surface is

$Q_1 = Q_{12} + Q_{13} =$ **22308 W**
$Q_2 = Q_{23} =$ **−3130 W**
$Q_3 = -Q_{23} - Q_{13} =$ **−19178 W**

so that surfaces 2, 3 are cooled. Note that the system is in thermal equilibrium so that $Q_1 + Q_2 + Q_3 = 0$. Inserting the values shows this condition to be satisfied.

The heat distribution (W) for each surface is shown in Fig. 7.21.

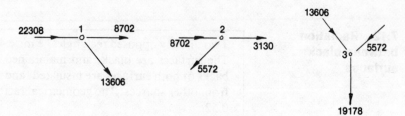

Figure 7.21

7.13 Radiation between three grey surfaces

An enclosure consists of three grey, opaque surfaces: a hot surface 1, cooler surface 2, and refractory surface 3.

Show that the net heat supply to surface 1 is given by

$$Q_1 = \frac{\sigma A_1(T_1^4 - T_2^4)}{\frac{1-e_1}{e_1} + \frac{A_1}{A_2}\frac{1-e_2}{e_2} + \left(F_{12} + \frac{A_2 F_{13} F_{23}}{A_1 F_{13} + A_2 F_{23}}\right)^{-1}}$$

A furnace of rectangular cross-section, 1 m height × 2 m width, is 2 m long. The roof is covered with electric heaters of rating 100 kW, emissivity 0.75. The floor is covered by ingots at 300 K, emissivity 0.40. The refractory side walls are completely insulated.
Geometrical factor: floor to roof = 0.42.
Calculate the heater temperature, wall temperature and net heat transfer between the roof and ingots.

Solution The equivalent electric circuit is shown in Fig. 7.22. In this problem surface 3 is refractory, meaning in this context that there is **no**

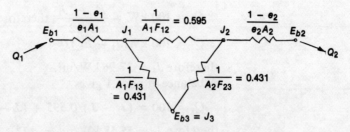

Figure 7.22

heat flow into or out of the surface. It therefore acts as a reflector and re-radiator: $Q_3 = 0$ so that $E_{b3} = J$. Also $Q_1 = -Q_2$.

The heat supplied to surface 1 is $Q_1 = (E_{b1} - E_{b2})/\text{total resistance}$. Now the total resistance between nodes 1 and 2 is made up of

$$\frac{1-e_1}{e_1 A_1}, \frac{1-e_2}{e_2 A_2}$$

and a resistance equivalent to the delta mesh of resistances $1/A_1 F_{12}$, $1/A_1 F_{13}$, $1/A_2 F_{23}$.

This equivalent resistance, say R_e, is equivalent to resistance $1/A_1 F_{12}$ in *parallel* with resistances $1/A_1 F_{13} + 1/A_2 F_{23}$. Hence

$$1/R_e = 1/A_1 F_{12} + 1/\left(\frac{1}{A_1 F_{13}} + \frac{1}{A_2 F_{23}}\right)$$

$$= \frac{1}{A_1 F_{12}}\left(\frac{1}{A_1 F_{13}} + \frac{1}{A_2 F_{23}}\right) \bigg/ \left(\frac{1}{A_1 F_{12}} + \frac{1}{A_1 F_{13}} + \frac{1}{A_2 F_{23}}\right)$$

Substituting $Q_1 = \dfrac{E_{b1} - E_{b2}}{\dfrac{1-e_1}{e_1 A_1} + R_e + \dfrac{1-e_2}{e_2 A_2}}$

which leads to the desired equation.

The numerical problem can be partially solved using the first equation. Thus

$$e_1 = 0.75 \qquad A_1 = 4 \text{ m}^2 \qquad F_{12} = 0.42$$
$$e_2 = 0.40 \qquad A_2 = 4 \text{ m}^2 \qquad F_{13} = F_{23} = 0.58$$
$$E_{b2} = 5.67 \times 10^{-8}(300)^4 = 0.459 \text{ kW/m}^2$$
$$Q_1 = 100 \text{ kW}$$

Substituting $100 = \dfrac{4(E_{b1} - 0.459)}{\dfrac{1}{3} + \dfrac{3}{2} + \left(0.42 + \dfrac{0.58 \times 0.58}{0.58 + 0.58}\right)^{-1}}$

giving $E_{b1} = 81.50 \text{ kW/m}^2$, therefore $81\,500 = 5.67 \times 10^{-8} T_1^4$ and $T_1 = $ **1094 K**.

The wall temperature T_3 and net heat transfer Q_{12} must be determined from the network.

$$Q_1 = 100 \text{ kW} = \frac{E_{b1} - J_1}{1/12}; \text{ therefore } J_1 = 73.17 \text{ kW/m}^2$$
$$Q_2 = 100 \text{ kW}(J_2 - E_{b2})/\tfrac{3}{8}$$

therefore $J_2 = 37.96$ kW/m².
A balance at node 1 gives

$$Q_1 = 100 = (J_1 - J_2)/0.595 + (J_1 - J_3)/0.431$$

therefore $J_3 = 55.58$ kW/m² $= \sigma T_3^4$
therefore $5.67 \times 10^{-8} T_3^4 = 55580$, giving $T_3 =$ **995 K**.
Finally, $Q_{12} = (J_1 - J_2)/0.595 =$ **59.15 kW**.

7.14 Radiation in gases

Two large grey plates at 800 K (emissivity 0.6) and 600 K (emissivity 0.2) are separated by a grey gas, of emissivity 0.10.

Calculate the net heat exchange between the surfaces/m², and the gas temperature.

Calculate the net heat exchange between the surfaces if the space between them is empty.

Solution The electrical analogy is shown in Fig. 7.23.

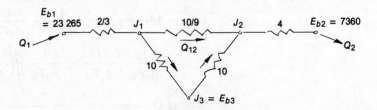

Figure 7.23

$$E_{b1} = 5.68 \times 10^{-8}(800)^4 = 23\,625 \text{ W/m}^2$$
$$E_{b2} = 5.68 \times 10^{-8}(600)^4 = 7360 \text{ W/m}^2$$

The various resistances are

$$\frac{1-e_1}{e_1} = \frac{0.4}{0.6} = \tfrac{2}{3}, \quad \frac{1-e_2}{e_2} = \frac{0.8}{0.2} = 4,$$

$$\frac{1}{\tau_g} = \frac{1}{1-e_g} = \frac{10}{9}, \frac{1}{e_g} = \frac{1}{0.1} = 10$$

The area is not included since it is taken as 1 m² for each plate. The geometrical factor between the plates is 1, neglecting end effects.

$$Q_1 = (E_{b1} - J_1)/(\tfrac{2}{3}) = (J_2 - E_{b2})/4$$

therefore, $\tfrac{3}{2}J_1 + \tfrac{1}{4}J_2 = 36\,738$.
A balance at node 1 gives

$\frac{3}{2}(E_{b1} - J_1) = \frac{9}{10}(J_1 - J_2) + \frac{1}{10}(J_1 - J_3)$

therefore, $34\,898 = \frac{5}{2}J_1 - \frac{9}{10}J_2 - \frac{1}{10}J_3$ and at node 3

$\frac{1}{10}(J_1 - J_3) = \frac{1}{10}(J_3 - J_2)$

therefore, $J_3 = \frac{1}{2}(J_1 + J_2)$.
Solving $J_1 = 21\,243$, $J_2 = 18\,054$, $J_3 = 19\,649$ W/m².
Hence the net exchange between the surfaces is

$Q_{12} = \frac{9}{10}(J_1 - J_2) = \mathbf{2870\ W/m^2}$

and $E_{bg} = 5.68 \times 10^{-8} T_g^4 = J_3 = 19\,649$; therefore $T_g = \mathbf{767\ K}$.
With no gas between the plates

$Q_{12} = (E_{b1} - E_{b2}) / \left[\dfrac{1 - e_1}{e_1 A_1} + \dfrac{1}{A_1 F_{12}} + \dfrac{1 - e_2}{e_2 A_2} \right]$

$= (23\,265 - 7360)/(\tfrac{2}{3} + 1 + 4) = \mathbf{2806\ W/m^2}$.

7.15 Temperature measurement

A thermocouple is inserted into a gas stream. The gas temperature is T_g, and the walls of the duct conveying the gas are at a temperature T_w. Derive an expression for the temperature recorded by the thermocouple, T_i.

Briefly outline how the difference between the temperatures T_i and T_g can be minimized.

Air at 1 atm flows along a duct at a velocity of 0.3 m/s. The duct wall temperature = 670 K. Diameter of the couple junction = 1.3 mm, emissivity = 0.2. The couple indicates a temperature of 830 K. For the flow across the wires $Nu_D = 0.5 Re_D^{0.5}$. The air properties are: $K = 0.06$ W/m K, $\nu = 0.85 \times 10^{-4}$ m²/s.

Calculate the convection and radiation heat transfer coefficients, and hence the true gas temperature.

Solution A heat-transfer balance on the thermocouple gives

convection (gas to thermocouple)
= radiation (thermocouple to wall)

therefore, $h_c A(T_g - T_i) = h_r A(T_i - T_w)$

where h_c = convection coefficient
h_r = radiation coefficient = $\sigma e (T_i^4 - T_w^4)/(T_i - T_w)$

Hence $T_g = T_i + (T_i - T_w) h_r / h_c$.

The difference $T_g - T_i = (T_i - T_w) h_r / h_c$ can be minimized by various means, such as

(a) lagging the duct so that $T_i - T_w$ is small,
(b) increasing the convection coefficient h_c. This can be achieved by using wires of a very small diameter, or by increasing the flow velocity of

the gas. The latter is the basis of the suction pyrometer, in which the gas velocity is increased and the indicated temperature T_i is recorded with the gas velocity. A plot of T_i against gas velocity gives a curve which flattens off, with increasing velocity, as T_i approaches T_g.

(c) by decreasing the radiation coefficient h_r. This can be done by putting a radiation shield around the thermocouple.

$Nu_D = 0.5 Re_D^{0.5}$ where

$$Nu_D = h_c D/K = h_c(0.0013)/0.06$$

and

$$Re_D = vD/\nu = 0.03(0.0013)/(0.85 \times 10^{-4}) = 45\,882$$

therefore

$$0.021\,67 h_c = 0.5(45\,882)^{0.5} = 214.2$$

giving $h_c = $ **9885 W/m² K**
$h_r = 5.67 \times 10^{-8} \times 0.2(830^4 - 670^4)/(830 - 670)$
$\nu = 19.35$ W/m² K
and $T_g = 830 + (830 - 670)(19.35)/9885 = $ **830.3 K**.

Problems

1 A plane slab of thickness 40 cm has the surface temperatures maintained at 100 °C and 200 °C. The cross-sectional area is 1.8 m². $K = 2.2$ W/m K.

Calculate the heat transmission rate.

Given that the thermal conductivity varies with the temperature T (K) according to the equation $K = 2.2 + 4 \times 10^{-4} T$ W/m K, determine the heat transmission rate.

Answer   990, 1066 W

2 A composite wall consists of three materials as shown in Fig. 7.24.

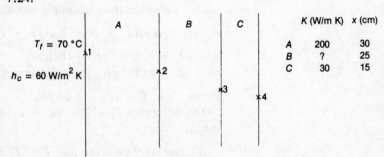

Figure 7.24

The surface temperatures are $T_1 = 50$ °C, $T_4 = 0$ °C.
Determine the conductivity of material B, and the interface temperatures.
Answer 7.11 W/m K; 48.2, 6 °C

3 The thermal conductivity of the insulation = 0.5 W/m K. If the surface temperature of the insulation $T_s \not> 50$ °C, calculate the surface coefficient and heat transfer rate.

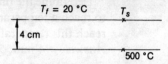

Figure 7.25

Answer 187.5 W/m² K; 5625 W/m²

4 A pipe of outside diameter 10 cm conveys steam at 120 °C. It is covered with insulation ($K = 0.15$ W/m K), 5 cm thick. The outer surface temperature of the insulation is 35 °C. Calculate the heat transfer rate/m length of pipe.

If the ambient air temperature = 15 °C determine the surface coefficient.
Answer 115.6 W/m; 9.2 W/m² K

Assuming the relationship $Nu_D = c(Gr\ Pr)^{1/4}$ for the surface coefficient, determine the value of the constant c.

$$Gr_D = \frac{\beta g \rho^2 D(T_s - T_f)}{\mu^2} \quad \text{where } \beta = 0.0034$$

Answer 1.12

5 A tank consists of a cylindrical centre section with two hemispherical ends. The tank contains a fluid that maintains the inside surface temperature of the tank at 350 °C. Tank thickness = 2.5 cm. The outer diameter of the cylinder is 2 m, and the length = 2 m. The ambient air temperature = 25 °C.

Convection coefficient (air-tank) = 7 W/m² K.

Calculate the heat that must be added to the fluid, and the surface temperature.

The thermal resistance for a sphere = $\dfrac{r_o - r_i}{4\pi K r_o r_i}$.

K (steel) = 14 W/m K.
Answer 56.5 kW, 346 °C

6 A large steel plate, 250 mm thick, is initially at 45 °C. One surface is exposed to air at 200 °C. The convection coefficient = 210 W/m² K.

For the plate $\alpha = 2 \times 10^{-6}$ m²/s, $K = 10$ W/m K.
Determine the surface temperature, and the temperature at a depth of 20 mm, after 180 s of exposure to the warm air.
Answer 97, 65 °C

7 A thick flat plate is initially at 50 °C. The surface is suddenly raised to 80 °C by hot gas. $\alpha = 2 \times 10^{-6}$ m²/s, $K = 10$ W/m K. The convection coefficient = 120 W/m² K.
Determine the time required for a plane 30 mm below the surface to reach a temperature of 65 °C, and the heat into the plane/m² to reach this temperature.
Answer 74 min, 5.1 MJ

8 The temperature profile along a simple rectangular fin is given by $\theta = \theta_0 \cosh m(L - x)/\cosh mL$, where $m = \sqrt{\dfrac{hP}{kA}}$, θ = local temperature difference.
Show that the fin efficiency is tanh mL/mL.
The fuel elements in a nuclear reactor are contained in cylindrical tubes with external longitudinal fins.
Fuel element can diameter = 40 mm.
Number of fins around the circumference = 24, fin thickness = 1.2 mm, fin height = 25 mm.
Can surface temperature = 500 °C, and cooling gas temperature = 420 °C. Thermal conductivity of the can and fin material = 150 W/m K. Convection coefficient (can to gas) = 700 W/m² K.
Determine the *total* convection heat loss/m length of fuel element.
Answer 36.4 kW

9 Define the Biot and Fourier numbers, and explain their significance in the solution of transient heat conduction problems.
A furnace wall, made of fireclay bricks, is 300 mm thick. The wall is initially at 20 °C. The furnace is started up by circulating hot gas at 300 °C within the furnace.
Neglecting convection between the gas and wall, determine the time taken for the midpoint of the wall to reach a temperature of 60 °C, and the total heat flow into the wall/m² up to that time.
Fireclay: $C_p = 0.962$ kJ/kg K, density = 2323 kg/m³, thermal conductivity = 1.367 W/m K.
Error function tables are provided.
Answer 142 min, 0.55 MJ

10 Show that the local temperature difference, θ, varies with position, x, along a simple extended surface according to the equation

$$\frac{d^2\theta}{dx^2} - \frac{h_c P}{KA}\theta = 0$$

A fin of square cross-section, 1 cm side, projects 15 cm from the surface to which it is attached. The surroundings are at 15 °C. Fin base temperature = 75 °C. $K = 120$ W/m K. Convection coefficient $h_c = 30$ W/m² K.

Determine the total heat loss rate from the fin and the proportion of this which occurs from the tip.

It may be assumed that the solution of the differential equation is $\theta/\theta_0 = \cosh[m(L-x)]/\cosh mL$, where $m = \sqrt{hP/kA}$.
Answer 7.9 W; none − insulated tip

11 Outline the process of heat convection from a hot plate to a fluid in contact with it, indicating the fluid properties that would be expected to influence the process.

A liquid metal flows at a rate of 3 kg/s through a tube of constant heat flux, 5 cm I.D. The fluid at 200 °C is to be heated with the tube wall 30 °C above the fluid temperature.

Calculate the tube length required for a 1 °C rise in the fluid bulk temperature.

$\rho = 7700$ kg/m³, $\nu = 8 \times 10^{-8}$ m²/s, $C_p = 130$ J/kg K, $K = 12$ W/m K.

$Nu_D = 0.625 Pe^{0.4}$ for $100 < Pe < 10^4$, $\frac{L}{D} > 60$.

(*Note:* Pe = Peclet number = $Re\, Pr$.)
Answer 3.07 cm

12 Air at 1 bar, 120 °C enters a pipe of I.D. 7 cm at a flow rate of 0.22 m³/s. The pipe wall temperature is constant at 20 °C. Determine the heat removed from the air, and the pipe length required, if the air is cooled to 80 °C.

$Nu_D = 0.023 Re_D^{0.8} Pr^{0.4}$.
Answer 7.88 kW, 3.44 m

13 A spherical storage tank is 5.0 m diameter and contains liquefied natural gas. The temperature of the outer surface is 10 °C. The ambient temperature is 20 °C. Surface emissivity = 0.4. Calculate the heat transfer rate to the gas when the wind velocity = 8 m/s.

For flow over spheres:

$$Nu_D = 2 + (0.4 Re_D^{1/2} + 0.06 Re_D^{2/3}) Pr^{0.4} \left(\frac{\mu_\infty}{\mu_s}\right)^{1/4},$$

where all properties of the air (except μ_s) are evaluated at the free-stream temperature μ_∞.
Answer 8.15 kW

14 A vertical pipe conveying steam passes through a room of volume 900 m³, whose exterior surface is completely insulated. Still air temperature = 20 °C. Exterior surface (pipe) temperature = 60 °C. O.D. of pipe = 6.8 cm, length = 3.4 m.

Determine the heat-transfer rate by convection from the pipe and the time taken to raise the room air temperature by 1 °C.
$Nu_D = c(Gr_D \, Pr)^n$.

Gr Pr	10^4–10^9	10^9–10^{13}
c	0.59	0.021
n	0.25	0.40

$$Gr_D = \frac{\beta g \rho^2 D^3 (T_s - T_\infty)}{\mu^2}$$

Answer 225 W; 76 min

15 Air flows over brass tubes containing steam. The surface convection coefficients on the air and steam sides of the tubes are 70 and 210 W/m² K respectively. The tubes are 1.8 cm I.D. and 2.1 cm O.D.

Calculate the overall heat transfer coefficient based

(a) on the inner area of the tube,
(b) on the outer area of the tube.

Determine the overall coefficient (a) if there is a fouling factor = 0.000 10 m² K/W on the inner surface.

Briefly outline the factors which determine the value of the overall coefficient.

Answer (a) 58.8, (b) 50.4; 58.4 W/m² K

16 Water enters a heat exchanger at 40 °C, and a flow rate of 12 kg/s. It is heated by hot air, flow rate 2 kg/s, entering at 460 °C.

The overall coefficient is 275 W/m² K. Surface area = 14 m².

Determine the water temperature at outlet and exchanger effectiveness if it is (a) a parallel flow, (b) a contraflow type.

Answer (a) 54 °C, 81 %; (b) 54.4 °C, 84 %

17 Dry steam at 125 °C is condensed on the outside of tubes in a shell and tube exchanger.

Water enters the tubes at 35 °C, mass flow rate 1.3 kg/s. Steam flow rate = 2.5 kg/s. Overall coefficient = 165 W/m² K based on a surface area of 3.7 m².

Calculate the heat transfer rate to the water, and water temperature at outlet.

Answer 5470 kW, 44 °C

18 Two parallel block rectangles 5 m × 10 m are directly opposed to each other and spaced 5 m apart. Geometrical factor = 0.18. The surroundings are black at 0 K.

The temperatures of the surfaces are 2000 K and 1000 K.

Calculate the net heat transfer between the surfaces, and the net energy supply to each surface.

Given that the surroundings are replaced by a refractory surface, determine the temperature of the refractory surface.

Answer 7655, 37 195, 2325 kW; 44 850, −5330, −39 520 kW: 1707 K

19 A black solar collector, area 50 m², receives a radiant flux of 800 W/m² from the sun.

Assuming that the surroundings are equivalent to a black body at 30 °C, calculate the equilibrium temperature of the collector, and net exchange between the collector and the surroundings.

Estimate the collector temperature if the surface coefficient $h_c = 6$ W/m² K.

Answer 114 °C; 40 kW; 86 °C

20 An insulated black rectangular surface on a space vehicle is normal to the solar radiation. Assuming that the sun is a black body at 5550 K, the surroundings are equivalent to a black body at 0 K, and neglecting any incident energy from other sources, calculate the equilibrium surface temperature.

Diameter of sun = 1.39×10^6 K m. Distance of craft from the sun = 1.5×10^8 K m.

Answer 104 °C

21 An oven, with roof heaters, is used to heat grey sheets ($e = 0.40$). The sheets are at a temperature of 500 K.

Neglecting convection, calculate the heat supplied to the heater, and loss through the walls.

Geometrical factor, roof to sheets = 0.42.

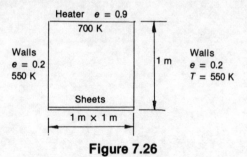

Figure 7.26

Answer 5.41, 2.72 kW

22 A cylindrical furnace is 50 cm diameter, 50 cm height. A heater, rating 1.3 kW, covers the floor. The stock to be heated covers the roof, and the walls are refractory. The emissivities of the surfaces are: heater 0.75; stock = 0.50.

The stock temperature = 500 K.

Neglecting conduction and convection, determine the temperature of the heater and wall, and net heat transfer to the stock. Geometrical factor (roof−floor) = 0.18.

Answer 803 K, 726 K; 1300 W

23 Two grey plates of emissivity 0.6 are separated by a distance of 4 cm. The space between the plates is filled with a grey gas, $e = 0.12$. The plate temperatures are 800 K and 600 K.

Calculate the net exchange between the plates, and net energy supplied to each plate

(a) with the gas filling the space,
(b) with no gas present.

Answer 6.62; 6.80 kW/m^2

8

Psychrometry: refrigeration, heat pumps

The energy consumption in the UK is spread amongst many sectors of industry, transport, commerce and domestic applications. The use of energy in commercial and domestic buildings for lighting, heating and air-conditioning amounts to a substantial proportion of the total energy consumption, and approximately one third is used to maintain a comfortable atmosphere in buildings. The factors contributing to human comfort are temperature, relative humidity and ventilation. Hence the study of psychrometry is important in the determination of a suitable environment.

In addition the principles of psychrometry are important in the areas of drying and industrial space heating. The field of space heating can also require refrigeration (cooling) and heat pump (heating) technology.

Psychrometry

Specific humidity ω = kg water vapour/kg dry air. Relative humidity ϕ = kg vapour per kg of air if saturated at the same temperature.

$$\omega = 0.622 p_v/p_a = 0.622 p_v/(p - p_v)$$

where p = total pressure; p_v, p_a = partial pressure of the vapour and air respectively.

$\phi = p_v/p_s$ where p_s = saturation pressure.

Enthalpy of moist air = $C_p T + \omega h_v$/kg air, where $C_p = 1.005$, T = dry bulb temperature, and h_v = enthalpy of vapour.

$21 < T < 65$ °C $\quad h_v = 2466.7 + 1.047 T$ kJ/kg K
$T < 21$ °C $\quad h_v = 2469.5 + 1.021 T$ kJ/kg K.

It is often convenient, and sufficiently accurate, to use a psychrometric chart: the enthalpy is read directly from the chart (Fig. 8.1).

It should be noted that, away from the critical point, the vapour can be treated as an ideal (perfect) gas and the equation $pv = mRT$ used.

Various units of pressure are used in psychrometric work. They are: the bar, pascal, mm mercury, and torr.

1 bar = 10^5 N/m^2 = 10^5 Pa, 1 atm = 101.3 kN/m^2
1 bar = 750 mm Hg, 1 atm = 760 mm/Hg
1 torr = 133.3 N/m^2 = 1 mm Hg
1 mm H$_2$O = 9.807 N/m^2

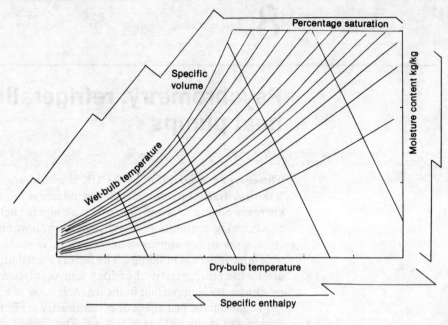

Figure 8.1

Psychrometric processes

The processes involved are heating, cooling of the air and water removal or addition. The application of the steady-flow equation to general processes (Fig. 8.2) gives

$$m_a h_{a2} + m_{v2} h_{v2} + m_{w1} h_f - Q_{12} = m_a h_{a1} + m_{v1} h_{v1}$$

and

$$m_a h_{a2} + m_{v2} h_{v2} + Q_{23} = m_a h_{a3} + m_v h_{v3} + m_{w2} h_f$$

where m_a = air-flow rate, m_v = vapour flow rate, and h_f = enthalpy of liquid water.

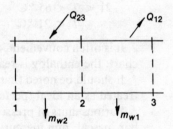

Figure 8.2

Refrigeration: heat pump

The simple vapour compression cycle was discussed in Chapter 3. In practice allowances must be made for pressure drops in the evaporator and condensor, undercooling, and compressor inefficiency.

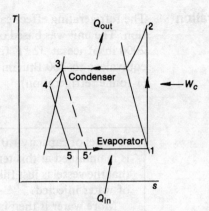

Figure 8.3

The effect of undercooling is shown in Fig. 8.3. The condensed vapour can be cooled to a temperature below the saturation temperature T_3. The refrigeration effect is then increased from $h_1 - h_5'$ to $h_1 - h_5$.

Flash chamber

In the throttling process, shown in Fig. 8.4 on the p–h diagram, some of the liquid flashes into vapour in passing through the valve. At any pressure the proportion of dry vapour to liquid $= ab/bc$. Since the dry vapour formed does not contribute to the refrigerating effect, it is more effective to bleed off the flash vapour and feed it back to the compressor.

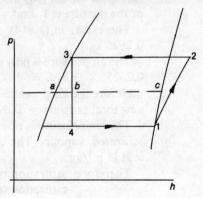

Figure 8.4

COP

The performance of a refrigerator is measured by the coefficient of performance, COP = refrigerating effect/compressor work = $(h_1 - h_5)/(h_2 - h_1)$.

In the case of a heat pump the same parameter can be used. However, since the purpose of the heat pump is to transfer heat to warm a space, rather than removing heat to cool a space,

$$COP = (h_2 - h_3)/(h_2 - h_1)$$

PSYCHROMETRY: REFRIGERATION, HEAT PUMPS 157

Unit of refrigeration

The refrigerating effect can be measured in terms of tonnes of refrigeration. The unit was based on the rate of heat removal required to produce 2000 lb of ice at 32 °F from water, at that temperature, in 24 h. It is equivalent to 200 Btu/min, which in the S.I. is equivalent to 3575 W (1 tonne refrigeration).

8.1 Basic psychrometry

A vessel of capacity 1.0 m³ contains air at 0.7 bar, 77 °C. The vessel is maintained at this temperature as water is injected into it. Given that the vessel is just filled with saturated vapour, calculate the mass of water injected.

More water is then injected until a *total* mass of 1.4 kg has been introduced. Calculate the total pressure in the vessel.

The vessel is now heated until all the water has just evaporated. Calculate the new total pressure, and the heat supplied.

Solution The calculations involve the use of the basic laws of psychrometry, and the thermodynamic properties of fluids tables.

At 77 °C, saturation pressure p_s = 0.42 bar, and the specific volume = 3.814 m³/kg. Hence the mass of vapour = 1/3.814 = 0.2622 kg.

Therefore, water injected = **0.2622 kg**. The total pressure at this condition = 0.7 + 0.42 = 1.12 bar.

When 1.4 kg water is injected it will be present in the vessel as dry saturated vapour (say m_s kg) and water (say m_w kg) such that the volume of the mixture is 1.0 m³.

Therefore, $m_s(3.814) + (1.4 - m_s)(0.001\,027) = 1$, giving m_s = 0.2625 kg.

The air pressure is now changed since the volume of saturated vapour = 0.2625 × 3.814 = 1.001 18 m³. The temperature has not changed, therefore the air pressure becomes 0.7 × 1.0/1.001 18 = 0.699 bar, and new total pressure = 0.699 + 0.42 = **1.119 bar**.

The temperature is now increased until the vessel is occupied by air + *saturated* vapour. The specific volume is therefore 1 m³/1.4 kg = 0.7143 m³/kg.

Therefore, saturation pressure = 2.517 bar.
saturation temperature = 127.6 °C.

The air pressure increases, as the temperature rises from 77 °C to 127.6 °C, to 0.7(127.6 + 273)/(77 + 273) = 0.801 bar, therefore total pressure = 2.517 + 0.801 = **3.318 bar**.

The heat supplied, Q, can be determined from an energy balance:

enthalpy, saturated vapour at 2.517 bar = 2717 kJ/kg
enthalpy, saturated vapour at 0.42 bar = 2638 kJ/kg
enthalpy, liquid vapour at 0.42 bar = 323 kJ/kg
mass of air = pv/RT
= 0.7 × 10⁵ × 1.0/(287 × 350) = 0.697 kg

Therefore, $Q = 1.4(2717) + 0.697(1.005)(127.6 - 77)$
$\qquad - 0.2622(2638) - (1.4 - 0.2625)(323)$
$\qquad = \mathbf{2780\ kJ}$.

8.2 Air cooling: dew point

> Air is supplied to a room at 17 °C with a relative humidity of 0.60. Barmetric pressure = 760 mm Hg. Calculate the specific humidity and determine the dew point.
>
> The air passes over a cooling coil at a rate of 1.0 m³/s, and leaves at 7 °C. Calculate the mass of vapour condensed, and the heat removal rate.

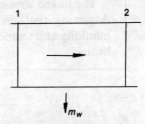

Figure 8.5

Solution At 17 °C, saturation pressure $p_{s1} = 0.019\,36$ bar. The partial pressure of the vapour is

$$p_{v1} = \phi_1 p_{s1} = 0.6 \times (0.019\,36) = 0.011\,62 \text{ bar}.$$

Therefore, specific humidity $\omega_1 = 0.622 p_{v1}/(p - p_{v1})$

$\qquad = 0.622 \times 0.011\,62/(1.013 - 0.011\,62) = \mathbf{0.0072\ kg/kg}$

If the air is cooled at constant pressure the vapour begins to condense at the saturation temperature corresponding to a pressure of 0.011 62 bar. From tables the temperature is approximately **9 °C**.

The air is now cooled to $T_2 = 7$ °C, which is below the dew point. Hence some vapour will condense to liquid, and the relative humidity at section 2 is $\phi_2 = 1$.

The air flow rate (constant) $m_a = p_a v/RT_a = (1.013 - 0.011\,62) \times 10^5 \times 1.0/(287 \times 290) = 1.203$ kg/s.

Hence m_{v1} = vapour flow rate at section 1
$\qquad = \omega_1 m_a = 0.0072 \times 1.203 = 0.008\,66$ kg/s

Now at section 2, $p_s = 0.010\,01$ bar (since $T_2 = 7$ °C), therefore, $\omega_2 = 0.622(0.010\,01)/(1.013 - 0.010\,01) = 0.0062$ kg/kg therefore, $m_{v2} = 0.0062 \times 1.203 = 0.007\,46$ kg/s giving vapour condensed = $m_{v1} - m_{v2} = \mathbf{0.0012\ kg/s}$.

The energy equation gives

$$m_a C_{pa}(T_1 - T_2) + m_{v1} h_{v1} - Q = m_{v2} h_{v2} + m_w h_f$$

where $h_{v1} = 2469.5 + 1.021(17) = 2486.9$ kJ/kg
$h_{v2} = 2469.5 + 1.021(7) = 2476.6$ kJ/kg
$h_f =$ (assuming removal at 7 °C) $= 29.4$ kJ/kg

Therefore, $Q = 1.203(1.005)(17 - 7) + 0.008\,66(2486.9)$
$\qquad\qquad - 0.007\,46(2476.6) - 0.0012(29.4)$
$\qquad = \mathbf{15.12\ kW}$ (removed).

8.3 Adiabatic mixing

4 m³/s of air at 22 °C, relative humidity 0.50, is mixed adiabatically with 2 m³/s of air at 3 °C, saturated. The pressure of each stream is 1 bar.

The mixed streams then pass through a heater, leaving at 10 °C. Assuming that no condensation occurs, determine the specific humidity and temperature after mixing, and the heat transfer in the heater.

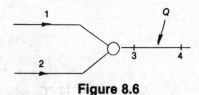

Figure 8.6

Solution For stream 1:
$T_1 = 22$ °C
therefore, $p_{s1} = 0.026\,42$ bar
$\qquad p_{v1} = \phi p_{s1} = 0.013\,21$ bar
therefore, $\omega_1 = 0.622 \times 0.013\,21/(1 - 0.013\,21)$
$\qquad\quad = 0.008\,33$ kg/kg

For stream 2:
$T_2 = 3$ °C
therefore, $p_{s2} = 0.007\,575$ bar
$\qquad p_{v2} = p_{s2}$ (saturated)
therefore, $\omega_2 = 0.622 \times 0.007\,575/(1 - 0.007\,575) = 0.004\,75$ kg/kg

For adiabatic mixing $m_1 h_1 + m_2 h_2 = (m_1 + m_2) h_3$.
Now $m_a = pv/RT$, therefore

$m_{a1} = (1 - 0.013\,21) \times 10^5 \times 4/(287 \times 295) = 4.662$ kg/s
$m_{a2} = (1 - 0.007\,575) \times 10^5 \times 2/(287 \times 276) = 2.506$ kg/s
$m_{v1} = \omega_1 m_{a1} = 0.008\,33 \times 4.662 = 0.0388$ kg/s
$m_{v2} = \omega_2 m_{a2} = 0.004\,75 \times 2.506 = 0.0119$ kg/s

and, $h_1 = 1.005(22) + 0.008\,33[2466.7 + 1.047(22)] = 42.85$ kJ/kg
$\qquad h_2 = 1.005(3) + 0.004\,75[2469.5 + 1.021(3)] \quad = 14.79$ kJ/kg

Substituting $4.662(42.85) + 2.506(14.79) = 7.168h_3$ giving $h_3 = 33.04$ kJ/kg.

Also, $\omega_3 = (0.0388 + 0.0119)/7.168 =$ **0.007 08 kg/kg**.

Now, $h_3 = 33.04$ kJ/kg $= 1.005 T_3 + 0.00708 h_{v3}$ where $h_{v3} = 2469.5 + 1.021 T_3$ if $T_3 < 21$ °C. Hence substituting gives $T_3 = 15.4$ °C.

The heat supplied is determined from a heat balance. There is no condensation in this case.

Therefore, $mh_4 = mh_3 + Q$

$$h_4 = 1.005(10) + 0.00708[2469.5 + 1.021(10)]$$
$$= 27.606 \text{ kJ/kg}$$
$$h_3 = 1.005(15.4) + 0.00708[2469.5 + 1.021(15.4)]$$
$$= 33.072 \text{ kJ/kg}$$

Therefore, $Q = 7.168(27.606 - 33.072) =$ **−39.18 kW** (extracted).

8.4 Cooling tower

Air enters the base of a cooling tower at 1 atm, 18 °C, relative humidity 0.6 at a rate of 11.0 m³/s. Water is sprayed into the tower at 40 °C, leaving at 20 °C. The air leaves the tower at 30 °C, saturated. Determine the mass flow rate of water, and the proportion of the water lost by evaporation.

If the air velocity leaving the tower is 2.5 m/s, determine the tower diameter (assuming a circular cross-section).

Solution The flow diagram is shown in Fig. 8.7.

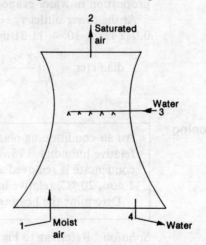

Figure 8.7

At section 1, $T_1 = 18$ °C, $\phi_1 = 0.6$, therefore,

$p_{s1} = 0.02063$ bar
$p_{v1} = 0.6 p_{s1}$
$\phantom{p_{v1}} = 0.01238$ bar

$$\omega_1 = 0.622 \times 0.012\,38/(1.013 - 0.012\,38)$$
$$= 0.007\,70 \text{ kg/kg}$$
$$m_{a1} = p_{a1}v_1/RT_1$$
$$= (1.013 - 0.012\,38) \times 10^5 \times 11/(287 \times 291)$$
$$= 13.18 \text{ kg/s}$$

At section 2, $T_2 = 30\,°C$, $\phi_2 = 1.0$, therefore

$$p_{s2} = p_{v2} = 0.042\,42 \text{ bar}$$
$$\omega_2 = 0.622 \times 0.042\,42/(1.013 - 0.042\,42)$$
$$= 0.027\,19 \text{ kg/kg}$$

An energy balance on the tower gives

$$m_a h_{a1} + m_{v1} h_{v1} + m_3 h_{f3} = m_a h_{a2} + m_{v2} h_{v2} + m_4 h_{f4}$$

where, $h_{a1} = 1.005(18) = 18.09$ kJ/kg
$h_{v1} = 2469.5 + 1.021(18) = 2487.88$
h_{f3} = liquid enthalpy at 40 °C = 167.50
$h_{a2} = 1.005(30) = 30.15$
$h_{v2} = 2466.7 + 1.047(30) = 2498.11$
h_{f4} = liquid enthalpy at 20 °C = 83.90
$m_{v1} = 13.18(0.007\,70) = 0.101\,49$ kg/s
$m_{v2} = 13.18(0.027\,19) = 0.035\,836$ kg/s

therefore water evaporated = $0.358\,36 - 0.101\,49 = 0.0256\,87$ kg/s.
Hence $m_4 = m_3 - 0.256\,87$ kg/s.

Substituting the values into the equation gives $m_3 = $ **9.33 kg/s** and the proportion of water evaporated = $0.256\,87/9.33 = $ **0.0275**.

At the tower outlet $v_a = m_a RT_2/p_{a2} = 13.18 \times 287 \times 303/(1.013 - 0.042\,42) \times 10^5 = 11.81$ m³/s, therefore

$$\text{diameter} = \sqrt{\left(\frac{4 \times 11.81}{\pi \times 2.5}\right)} = \mathbf{2.45 \text{ m}}$$

8.5 Air conditioning

An air conditioning plant takes in 2.5 m³/s of air at 1 atm, 25 °C, relative humidity 0.75. The air passes through a cooler, where the condensate is removed at 9 °C, and then heated before delivery at 1 atm, 20 °C, relative humidity 0.49.
Determine the heating and cooling loads.

Solution Referring to Fig. 8.8, $T_1 = 25\,°C$.

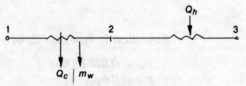

Figure 8.8

Therefore $p_{s1} = 0.03166$ bar
$\phi_1 = 0.75$
$p_{v1} = 0.02375$ bar
$\omega_1 = 0.622 \times 0.02375/(1.013 - 0.02375)$
$= 0.0149$ kg/kg
$T_3 = 20\ °C$
Therefore $p_{s3} = 0.02337$ bar
$p_{v3} = 0.49 p_{s3}$
$= 0.01145$ bar
$\omega_3 = 0.622 \times 0.01145/(1.013 - 0.01145)$
$= 0.0071$ kg/kg

There is no water removal in the heater and the air mass flow rate is constant, so that $\omega_2 = \omega_3$ and

$$m_{a1} = m_{a2} = m_{a3} = 0.98925 \times 10^5 \times 2.5/(287 \times 298) = 2.89\ \text{kg/s}.$$

At section 2, $\omega_2 = 0.0071$ kg/kg $= 0.622 p_{v2}/(0.013 - p_{v2})$; therefore $p_{v2} = 0.01143$ bar, and the air is saturated, therefore $p_{s2} = 0.01143$ bar and from tables $T_2 = 9\ °C$.

An energy balance on the cooler gives

$$Q_c = m_a(h_{a2} - h_{a1} + \omega_2 h_{v2} - \omega_1 h_{v1}) + m_w h_f$$

where $h_{v2} = 2469.5 + 1.021(9) = 2478.7$ kJ/kg
$h_{v1} = 2466.7 + 1.047(25) = 2492.9$ kJ/kg
$h_f = 37.8$ kJ/kg
$m_w = m_a(\omega_1 - \omega_2) = 2.89(0.0149 - 0.0071) = 0.0225$ kg/s

Therefore $Q_c = 2.89(1.005)(9 - 25) + 2.89(0.0071 \times 2478.7$
$- 0.0149 \times 2492.9) + 0.0225(37.8) = -\textbf{102.1 kW}$

and a balance on the heater gives

$$Q_h = m_a(h_3 - h_2 + \omega_3 h_{v3} - \omega_2 h_{v2})$$
$= 2.89[1.005(20 - 9) + 0.0071(2489.9 - 2478.7)]$
$= \textbf{32.2 kW}.$

8.6 Vapour compression cycle

The pressure in the evaporator of an ammonia refrigerator is 1.902 bar, and in the condensor 12.37 bar. Calculate the ideal coefficient of performance and refrigerating effect when working between the corresponding saturation temperatures.

Determine these values when the cycles are modified to

(a) dry saturated vapour delivered to the condensor,
(b) dry saturated vapour delivered to the compressor,
(c) as in (b) with undercooling by 11 °C.

Also calculate, for each cycle, the ammonia flow rate/tonne refrigeration (3575 W).

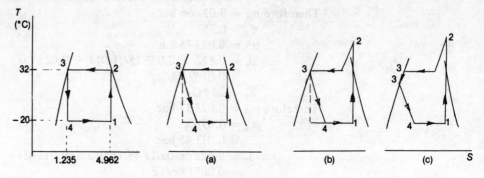

Figure 8.9

Solution In the ideal reversed Carnot cycle (Fig. 8.9)

$$COP = T_1/(T_2 - T_1) = 253/52 = 4.86$$

Refrigeration effect is

$$T_1(s_1 - s_4) = 253(4.962 - 1.235) = 943 \text{ kJ/kg}$$

(a) $h_2 = 1469.9$ kJ/kg, $s_2 = 4.962 = s_1 = 0.368 + 5.225 x_1$, therefore $x_1 = 0.874$ and

$h_1 = 89.8 + 0.874(1330.2) = 1252.4$ kJ/kg
$h_4 = h_3 = 332.8$ kJ/kg   (throttling process)
$COP = (h_1 - h_4)/(h_2 - h_1) = \mathbf{4.23}$
Refrigeration effect $= h_1 - h_4 = \mathbf{919.6 \text{ kJ/kg}}$

(b) At point 2, $s_2 = 5.623$ and the vapour is superheated by 84 °C to $T_2 = 116$ °C. $h_2 = 1699.0$ kJ/kg K. $h_1 = 1420.0$ kJ/kg K.

$COP = (1420.0 - 332.8)/(1699.0 - 1420.0) = \mathbf{3.90}$
Refrigeration effect $= h_1 - h_4 = \mathbf{1087.2 \text{ kJ/kg}}$

(c) In this case undercooling takes place to point $3'$. Assuming that it takes place along the saturated liquid line, $h'_3 = $ liquid enthalpy at $T_3 = 32 - 11 = 21$ °C, therefore

$h'_3 = 279.9$ kJ/kg $= h_4$
$COP = (1420.0 - 279.9)/(1699.0 - 1420.0) = \mathbf{4.09}$
Refrigeration effect $= h_1 - h_4 = \mathbf{1140.1 \text{ kJ/kg}}$

The ammonia flow rate $= 3575/$refr. effect, giving the four values as **3.79, 3.89, 3.29** and **3.14 kg/s** respectively.

8.7 Refrigeration cycle

A refrigerator using Freon-12 has an evaporator saturation temperature of -25 °C, and a condenser saturation temperature of 30 °C. The vapour is dry saturated at entry to the compressor, and leaves at 50 °C. There is no undercooling of the condenser liquid.

By use of the $p-h$ chart, estimate the coefficient of performance, the compressor isentropic efficiency, and power input.

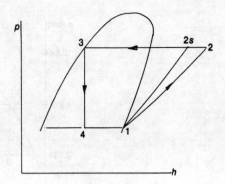

Figure 8.10

Solution The chart is illustrated in Fig. 8.10.

1–2s isentropic compression
2–3   constant pressure condensation
3–4   adiabatic throttling
4–1   constant enthalpy (adiabatic) throttling

Compressor work = $h_2 - h_1$ = 213 − 176 = 37 kJ/kg
Refrigeration effect = $h_1 - h_4$ = 176 − 65 = 111 kJ/kg
  Therefore, COP = 111/37 = **3.0**
Mass flow of refrigerant = 3575/111 × 10^3 = 0.032 kg/s and power input = 37 × 0.032 = **1.19 kW**.
  Compressor isentropic efficiency = (206 − 176)/(213 − 176)
                                    = **0.81**.

8.8 Flash chamber

A vapour-compression refrigerator uses Fr-12, with an evaporator pressure of 2.191 bar, and a condensor pressure of 10.84 bar. The vapour is dry saturated at the compressor inlet, and the condensate is undercooled by 10 °C. The compression is carried out in two stages and a flash chamber used at the interstage pressure of 4.914 bar.

Calculate the mass of vapour bled off in the flash chamber, the vapour condition at entry to the second compressor stage, refrigerating effect, and the coefficient of performance.

Solution Refer to Fig. 8.11, and consider 1 kg refrigerant flow through the condensor.
  T_1 = −10 °C, h_1 = 183.19 kJ/kg, s_1 = 0.7020.
  Since $s_2 = s_1$, at point 2 the vapour is superheated to approximately 20 °C and h_2 = 197.22 kJ/kg.

$$h_5 = h_6 = 79.71 \text{ kJ/kg}$$
$$h_7 = h_8 = 50.10 \text{ kJ/kg}$$

The mixture at point 6 consists of x kg dry vapour + $(1 - x)$ kg liquid, and $h_5 = h_{f7} + x h_{fg7}$.

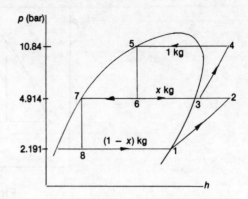

Figure 8.11

Therefore $x = (79.71 - 50.10)/143.68 = 0.206$, so bled vapour = **0.206 kg/kg**.

The mixing at the second compressor stage inlet is adiabatic, and x kg dry vapour at p_6 is mixed with $(1 - x)$ kg superheated vapour at p_6. The condition of the vapour at point 2 can be determined from the isentropic first-stage compression.

$s_1 = s_2 = 0.7020$, and therefore $h_2 = 197.22$ kJ/kg.

Hence, $h_3 = (1 - x)h_2 + xh_{g7} = 0.794(197.22) + 0.206(193.78)$
$= 196.57$ kJ/kg

Refrigeration effect $= (1 - x)(h_1 - h_8) = 0.794(183.19 - 50.10)$
$= 105.7$ kJ/kg

At point 4, $s_4 = s_3 = 0.6996$; therefore the vapour is superheated and $h_4 = 210.9$ kJ/kg.

Compressor work $= (1 - x)(h_2 - h_1) + (h_4 - h_3) = 25.47$ kJ/kg therefore $COP = 105.7/25.47 =$ **4.15**.

8.9 Heat pump

A heat pump, using ammonia, operates on a vapour-compression cycle. It is used to heat 0.5 m³/s air to 30 °C, from source air at 5 °C. The air pressure is constant at 1 atm. The temperature in the evaporator is -7 °C and the pressure in the condenser is 14.7 bar. The vapour is dry saturated at the compressor inlet, and liquid enters the throttle valve at 25 °C.

C_p (air) $= 1.005$ kJ/kg K.

Determine the refrigerant flow rate and power input to the compressor; and coefficient of performance.

Solution $s_2 = s_1 = 5.433$ therefore $h_2 = 1655.9$ kJ/kg.
The air flow rate is

$m_a = pv/RT$
$= 1.013 \times 10^5 \times 0.5/(287 \times 303)$
$= 0.582$ kg/s

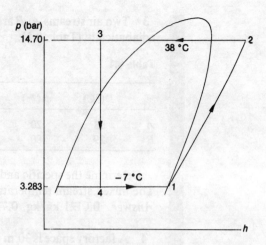

Figure 8.12

therefore heat required

$= 0.582(1.005)(30 - 5)$
$= 14.63 \text{ kW}$

The heat rejected to the heated space $= h_2 - h_3 = 1354.1 \text{ kJ/kg}$, therefore refrigerant flow rate $= 14.63/1354.1 = \mathbf{0.0108 \text{ kg/s}}$.

Compressor work $= h_2 - h_1 = 219.4 \text{ kJ/kg}$, and the power input $= 219.4 \times 0.0108 = \mathbf{2.37 \text{ kW}}$.

COP = heat rejected/compressor work = 1354.1/219.4
$= \mathbf{6.17}$.

Problems

1 Air, at 28 °C, relative humidity 0.80, is cooled at a constant pressure of 1 bar to 7 °C, and saturated. The initial flow rate = 16 800 m³/h. The condensed water vapour leaves the cooler at 4 °C.

Calculate the initial and final specific humidity, vapour condensation rate and heat transfer rate.

Answer   0.0194, 0.0063 kg/kg; 0.0686 kg/s; 278 kW

2 Air is drawn into an air conditioning system from outside conditions of 5 °C, relative humidity 25 %. It is heated by a heat supply of 1000 kW, and then humidified isothermally with a water evaporation rate of 1550 kg/h. The air is delivered to a machine shop, 40 m × 40 m × 10 m, and there are 10 complete air changes/hour. Air pressure = 1.0 bar.

Determine the final temperature of the air, and final humidity.

Answer   24 °C; 51 %

3 Two air streams A, B are mixed at a constant pressure of 1 atm, adiabatically (Table 8.1).

Table 8.1

	T (°C)	ϕ (%)	Q (m³/s)
A	13	20	0.40
B	29	80	0.50

Determine the specific and relative humidity of the mixed streams. The mixed stream temperature should also be estimated.
Answer 0.0121 kg/kg, 0.72; 22 °C

4 A factory space is 30 m × 10 m × 10 m, and has 8 complete air changes per hour. The air drawn in is at 27 °C, relative humidity 75 %. It passes over a cooling coil to become saturated, and then over a heating coil to enter the space at 18 °C, relative humidity 50 %. The heating coil is supplied with saturated steam at 1.4 bar.

Using the psychrometric chart estimate the air temperature between the two coils, the heat transfer to the cooler, the heat transfer from the heater, and flow rate of the heater steam.
Answer 8 °C; 370, 80 kW; 0.036 kg/s

5 A room is supplied by saturated air at 2 °C, and the room conditions are required to be 20 °C, relative humidity 50 %. The supply air is heated, and then water at 10 °C is sprayed into it to give the required humidity. The total pressure is constant at 1.02 bar.

Determine the temperature to which the air must be heated and kg spray water/m³ air at room conditions required.
Answer 27 °C; 0.0035

6 Air at 1.005 bar, 1 °C, relative humidity 95 % is to be conditioned for use in a room where the required conditions are 25 °C, relative humidity 45 %. The intake air flow is 10 m³/s.

Determine the heat supply rate prior to spraying in water at 6 °C, and the water flow rate to the sprays.
Answer 440 kW; 0.065 kg/s

7 A cooling tower is required to cool water from 50 °C to 25 °C at a water flow rate of 2 000 000 kg/h.

The inlet air conditions are 15 °C, $\phi = 70$ %. The air at outlet is saturated at 30 °C: and leaves the tower at a mean velocity of 5 m/s.

Calculate the water evaporation loss, and tower diameter.
Answer 3.0 %, 14.3 m

8 Briefly explain how the wet bulb temperature of an air stream is measured, and why it differs from the dry bulb temperature. An evaporative cooler is required to cool 2 250 000 kg/h of water from 32 °C to 18 °C.

The air at inlet is at 1 bar, 16 °C (dry bulb), 10 °C (wet bulb). The exit air is saturated at 3 °C.

Using the psychrometric chart, determine the mass flow rate of air entering the cooler and the water evaporation loss.
Answer   483 kg/s; 1.9 %

9 (a) Air at $15\frac{1}{2}$ °C (dry bulb), 10 °C (wet bulb) is humidified by the addition of dry saturated steam at $13\frac{1}{2}$ bar. The air temperature is raised to 17 °C. Determine the specific humidity of the heated air, assuming that it is saturated and kg steam added/kg air.

(b) Air at $4\frac{1}{2}$ °C (dry bulb), $1\frac{1}{2}$ °C (wet bulb) is heated, and then humidified, to 21 °C, relative humidity = 40 %. The water is supplied at 10 °C.

Determine the water mass added/kg air, and the heat supplied/kg air.

In both solutions the psychrometric chart should be used.
Answer   (a) 0.012 kg/kg, 0.01 kg/kg, (b) 0.003 kg, 24.4 kJ

10 A dehumidifier comprises a vapour compression refrigeration unit with the evaporator installed in an air duct. Air at 35 °C, relative humidity 0.7, is drawn into the duct at a rate of 25 kg/s. The cooled air leaves the evaporator coil, saturated, at 15 °C.

In the Freon-12 cycle, dry saturated vapour leaves the evaporator at 3.08 bar, and is compressed to 12.19 bar (compressor isentropic efficiency = 0.70). Heat from the condensor is rejected to the ambient air. Assuming that there is no undercooling, determine the mass of water removed, the rate of heat removal in the evaporator, the refrigerant flow rate/kg air cooled, and compressor power.
Answer   0.37 kg/s; 1428 kW; 13.9 kg/s; 540 kW

11 A building space is to be maintained at 25 °C (dry bulb), relative humidity 0.5 when the ambient air is at 1 °C (dry bulb), −1 °C (wet bulb). The ambient air is preheated to 10 °C before being mixed with recirculated air in the ratio 4 kg recirculated air/kg ambient air.

The mixed air is then heated to 34 °C (dry bulb), and moisture is added. The humidification process is along a line of constant wet-bulb temperature.

Further sensible heating occurs before the air passes through the fan (giving a rise of 1 °C in the air temperature), and finally the air is supplied to the building at 33 °C (dry bulb), relative humidity 0.4. The volume of air entering the building is 8 m³/s.

Determine the sensible and latent heat gains or losses in the building, the mass of water added in the humidification process, and fan power required.
Answer   99.4, 40.7 kW loss; 0.041 kg/s; 9.0 kW

12 A room is maintained at 20 °C by air supplied at a rate of 1.5 kg/s, drawn from a mixing unit. The mixing unit takes air from two ducts, one carrying cold air and the other warm air, at 1 atm.

The room has a sensible heat loss of 14.7 kW and a latent heat gain of 3 kW. The supply air is made up of 25 % by mass dry air from the cold duct at 10 °C, specific humidity 0.003 kg/kg, and the remainder from the warm duct air, specific humidity 0.01 kg/kg.

Using the psychrometric chart, determine the warm air duct temperature, volume flow rate in the warm air duct, and relative humidity in the room.
Answer 31 °C, 1.10 m^3/s, 0.59

13 A vapour-compression refrigerator using Freon-12 has a two-cylinder single-acting compressor, 80 mm bore, 110 mm stroke. Compressor speed = 600 rev/min, isentropic efficiency = 0.8, volumetric efficiency = 0.75. The evaporator and condensor pressures are 2.191 and 9.607 bar respectively. The vapour is dry saturated at the compressor entry, and is saturated liquid at the throttle valve entry.

The average refrigerating effect of 3.5 kW is obtained by intermittent operation of the plant. Determine the ratio of the operating time to total time, and power consumption when the unit is running.
Answer 0.30; 1.06 kW

14 A heat pump uses R12 as refrigerant without subcooling. The evaporator and condensor temperatures are 5 °C and 50 °C respectively. Vapour compression takes place from the saturated vapour state, and compressor isentropic efficiency = 0.80.

Determine the *COP* of the heat pump. Assuming the efficiency of the electric motor drive is 0.9, calculate the overall *COP*.

Discuss the merits and disadvantages of using the electrically driven heat pump compared to direct resistance heating for space heating.
Answer 3.87, 3.48

15 A vapour-compression air-to-air heat pump is to be used for domestic air conditioning. The plant uses Freon-12. For winter use the vapour is dry saturated at the compressor entry and the condensate is at the saturation temperature at the condensor outlet. There is a 20 K temperature difference between the circulating air and the Freon leaving the condensor, and a 10 K difference between the source air and Freon in the evaporator. For all conditions the compressor volumetric efficiency and speed are constant, and the isentropic efficiency = 0.72.

The unit is to be used for a house in which, for an indoor temperature of 20 °C when the outside temperature is 0 °C, the heat loss is 12 kW. Given that the heat pump is sized to supply the entire heating load down to an outside temperature of 5 °C, determine the Freon mass flow rate for this condition.

Calculate the power input to the compressor and the auxiliary heating required to cope with an outside temperature of −15 °C.
Answer   0.063 kg/s, 2.02 kW; 9 kW

16 A heat pump using Freon-12 is to provide domestic space heating for a maximum demand of 20 kW. The source heat is waste water at 10 °C. The circulating water of the heating system is to be raised to 75 °C.

Assuming that there is a 10 K temperature difference between fluids in both heat exchangers, compressor isentropic efficiency = 0.8, the Freon enters the compressor dry, saturated, and there is no undercooling, determine the *COP* and compressor power input.
Answer   1.32; 8.6 kW

9

Alternative (renewable) sources of energy

There are many sources of energy available, which can be used as alternatives to the conventional sources of fossil fuels (including nuclear energy). They are of importance in developing countries where fossil fuels are not available or must be imported at high costs, and will probably assume some importance in industrial countries as the cost increases and, in the long term, their availability decreases. The high energy requirements of the more affluent areas of the world can create difficult economic, social and environmental problems: for example, the oil market influenced by the 1973 price rise and formation of OPEC, the concern about radioactive wastes, and the high production of CO_2, SO_2, NO_x and unburnt hydrocarbons.

The use of energy is a key to food supply, to physical comfort, to quality of life, and even though the use of alternative sources of energy may be limited to small-scale applications, such usage can improve the life of small communities.

The alternative sources available for electrical power and/or heat generation can be summarized as follows:

Solar energy	• turbine generators (hydroelectric)
	• ocean thermal energy
	• solar satellites
	• photovoltaic devices
	• thermoelectric generators
	• collectors
Wind energy	• windmills (aero generators)
Tidal energy	• barrages
Wave energy	• wave generators
Heat	• geothermal
Chemical energy	• fuel cells
	• biomass
Non-fossil fuels	• refuse incineration
	• refuse pyrolysis

These alternative sources can also be described as renewable in the sense that the natural forces which produce them are virtually unchanging and inexhaustible.

Solar energy

The total amount of solar energy reaching the earth is enormous, something of the order of 4×10^{15} kWh per day. The amount reaching the earth's surface is dependent upon the ozone layer, dust, clouds and latitude. It is widely dispersed, and is readily convertible to low-grade heat.

The solar energy can be used to produce hot water, for space heating, drying by the use of fairly simple collectors, and by using concentrators to produce steam and electric power.

Photovoltaic cells use the properties of semiconductors to convert solar energy into electricity, but are very expensive. Their efficiency and size can be improved by the use of concentrators, and the use of concentrators also makes it possible to use thermionic or thermoelectric systems. An economically viable solar cell may be achieved in time.

Wind power

The use of wind power has been practised for many years in man's history, and is at present an economic technology for power generation and pumping. The difficulties involved in the conversion include the size of the windmill, the unpredictable variation in the wind velocity, direction and duration, the need to cater for extreme wind speeds, and the noise generated. An empirical law for the power output is

$$P \text{ (kW)} = 0.37 A (v/10)^2$$

where A = area swept out by the blades (m^2), and v = wind speed (m/s).

Tidal energy

Tidal mills were used in the eleventh century, but like many renewable sources, were displaced by cheap, readily available fossil fuels.

Tidal power is used by placing a barrage across a suitable bay or estuary to contain the sea water as the tide rises. The water is then discharged, at low tide, through turbines in the barrage to generate power. The maximum electrical energy that can be generated during a tidal cycle $= \eta \rho g R^2 S$, where η = conversion efficiency, R = tidal range, S = surface area of tidal basin.

There are no waste products or pollution, and the energy is reasonably consistent; but the number of suitable sites is limited, the barrage construction can be expensive, sea water can cause corrosion, and there is an effect on the environment due to the changes in the tidal current pattern.

Wave energy

Wave energy is most available during winter months, and can be substantial. The power available from waves off the UK coast can reach 80 kW/m. Problems in the conversion of wave energy include the variability of waves in height, wavelength, length, duration, the hostile marine environment, and the cost of transmitting the power developed to the shore.

Geothermal energy

The total amount of heat stored in the earth is vast, but it can only be exploited in areas having suitable geological formations. The deposits may occur as dry steam, wet steam or hot water, and they have been used extensively in some countries for various purposes such as district heating, hot-water supply, horticulture and electricity generation.

Problems of using those fields that are available include drilling costs, reliability of downhole pumps, and corrosion due to mineral salts.

Fuel cells

A fuel cell provides a means for generating electricity without the Carnot limitation, achieving an efficiency of 60 %.

The cell process is essentially the reverse of electrolysis, where hydrogen (or similar gaseous fuel) and oxygen are introduced at the negative and positive electrodes respectively. The ensuing ionization produces electrons.

Fuel cells are attractive as peaking plant because of their ease of operation and maintenance of their efficiency at low output.

Non-fossil fuels

Biomass energy can be used by combustion of the plants for heating or power generation, anaerobic digestion to yield methane, fermentation to produce ethanol, or pyrolysis to give methanol.

The waste products (refuse) can also be burned dirctly in boiler plant to raise steam, or converted into pellets of refuse-derived fuel (RDF), or pyrolysed to liquid/gaseous fuels.

One scheme in use (in Brazil) is to convert biomass into methanol or ethanol. The latter is produced by the hydrolysis of cellulose or starch to produce sugar, followed by fermentation to produce alcohol.

$$C_6H_{10}O_5 + H_2O \rightarrow C_6H_{12}O_6$$
$$\text{cellulose} \qquad\qquad \text{sugar}$$

$$C_6H_{10}O_6 \rightarrow 2CH_3CH_2OH + 2CO_2$$
$$\text{sugar} \qquad \text{alcohol}$$

9.1 Solar radiation

Outline the reasons why the intensity of solar radiation is reduced as it passes through the earth's atmosphere.

Estimate the effective temperature of the earth T_e (assuming that the earth radiates as a black body). Take the reflectivity of the earth's surface = $\frac{1}{3}$, the radius of the sun = 7×10^5 km, temperature of sun (assumed to be a black body) = 6000 K, the distance between the sun and the earth = 1.5×10^8 km, and the radius of the earth = 6000 km.

Comment on the value obtained.

Solution The intensity is reduced by

(a) Dry air molecular absorption and scattering (Rayleigh scattering). The theoretical amount of radiation scattered by the particles is proportional to $1/\lambda^4$, where λ = wavelength. Thus short-wavelength (blue) light is more effectively scattered than longer-wavelength (red) light.

(b) Absorption and scattering from dust.

(c) Selective absorption by water vapour, CO, CO_2. The visible region contains few absorption bands, the main absorption being in the ultra-violet and infra-red regions. Thermal radiation emitted from a surface is between wavelengths of 10^{-7} and 10^{-4} m (0.1 to 100 μm), visible radiation is in the range 0.38 to 0.75 μm. The absorption by CO_2 and H_2O is strong in the infra-red, up to about 7 μm; and occurs in the lower atmosphere (below about 50 km).

(d) Reflection and absorption in cloud layers. The presence of large particles in clouds produces greater scattering of all light giving a sky which is much less blue, and eventually producing the familiar white appearance.

Radiation emitted by the sun $= \sigma 4\pi R_s^2 T_s^4$, and as it travels outwards through a distance R the energy/m^2 is

$$S = \sigma(R_s/R)^2 T_s^4 = (7 \times 10^5/1.5 \times 10^8)^2 (6000)^4 (5.67 \times 10^{-8})$$
$$= 1600 \text{ W/m}^2.$$

The total emission from the sun $= 4\pi R^2 S = 4.5 \times 10^{26}$ W, of which the earth intercepts $\pi R_e^2 S = \pi (6 \times 10^6)^2 (1600)$ W $= 1.8 \times 10^{14}$ kW.

Now the effective temperature of the earth, T_e, can be estimated by a balance between the incident solar radiation and outgoing reflection.

$$\pi R_e^2 S (1 - \alpha) = 4\pi R_e^2 \sigma T_e^4$$

where α = absorptivity.

Therefore $T_e = \left[\dfrac{(1-\alpha)S}{4\sigma} \right]^{1/4} = 262$ K or -11 °C.

This value is an *effective* temperature for the earth-atmosphere system. In actuality some of the radiation emitted from the earth's surface is absorbed in the atmosphere, and this is re-radiated in all directions. Hence some of it is radiated back towards the earth's surface and the temperature is greater than 262 K – the so-called 'greenhouse effect'.

9.2 Flat plate collector

A flat plate collector is shown in Fig. 9.1.
Temperatures:

T_1 – top cover
T_2 – lower cover
T_c – collector plate
T_a – ambient

Derive an expression for the collector efficiency.

Solution The heat collected by the working fluid in the pipes is $Q_u = mC_p(T_{fo} - T_{fi})$, where m = mass flow rate of fluid, T_{fo} and T_{fi} = fluid temperature at outlet and inlet respectively.

An energy balance on the collector, area A_c, gives

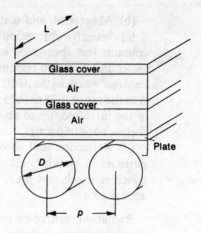

Figure 9.1

$$I_c A_c \tau_s \alpha_{s,c} = Q_u + Q'$$

where I_c = solar radiation incident on collector surface
τ_s = effective solar transmittance of collector covers
$\alpha_{s,c}$ = solar absorptivity of collector absorber plate surface
Q' = loss from plate to surroundings.

The instantaneous efficiency $\eta_c = Q_u/A_c I_c$.

The loss can be expressed in terms of an overall coefficient, U_c,

$$Q' = U_c A_c (T_c - T_a)$$

where T_c = mean plate temperature
T_a = ambient temperature.

The heat transfer/m² collector area between the plate and second glass cover is

$$Q_2' = A_c h_{c2}(T_c - T_2) + \sigma A_c (T_c^4 - T_2^4) / \left(\frac{1}{e_p} + \frac{1}{e_2} - 1 \right)$$

where h_c = heat transfer coefficient between plate and second glass cover
e_p = plate emittance
e_2 = cover emittance.

This can be put into a more convenient form by linearizing the radiation term,

$$T_c^4 - T_2^4 = (T_c^2 - T_2^2)(T_c^2 + T_2^2)$$
$$= (T_c - T_2)(T_c + T_2)(T_c^2 + T_2^2)$$

Hence

$$Q_2' = (h_{c2} + h_{r2}) A_c (T_c - T_2) = (T_c - T_2)/R_3$$

where

176 SOLVING PROBLEMS IN APPLIED THERMODYNAMICS AND ENERGY CONVERSION

$$h_{r2} = \sigma(T_c + T_2)(T_c^2 + T_2^2)/\left(\frac{1}{e_p} + \frac{1}{e_2} - 1\right)$$

Similarly the heat transfer between the two cover plates is

$$Q_1' = (h_{c1} + h_{r1})A_c(T_2 - T_1) = (T_2 - T_1)/R_4$$

where

$$h_{r1} = \sigma(T_1 + T_2)(T_1^2 + T_2^2)/\left(\frac{1}{e_1} + \frac{1}{e_2} - 1\right)$$

Between the top cover and the sky the loss is

$$A_c(h_{ca} + h_{ra})(T_1 - T_a) = (T_1 - T_a)/R_5$$

where

$$h_{ra} = e_1\sigma(T_1 + T_\infty)(T_1^2 + T_\infty^2)(T_1 - T_\infty)/(T_1 - T_a)$$

The temperature T_∞ = effective sky temperature, which is not equal to T_a (see worked example). Radiation occurs between the top cover and the sky, and convection between the cover and ambient air.

The total resistance can now be obtained

$$U_c = \frac{1}{R_1} + \frac{1}{R_3 + R_4 + R_5}$$

where R_1 = thermal resistance of the collector bottom.
Hence $\eta_c = Q_u/A_cI_c = (A_cI_c\tau_s\alpha_{sc} - Q')/A_cI_c$
$= \tau_s\alpha_{sc} - U_c(T_c - T_a)/I_c$

9.3 Collector calculation

A flat plate collector has two glass cover plates.
Collector area = 5 m²
Solar incident radiation = 800 W/m²
Ambient air temperature = 10 °C, effective sky temperature = 0 °C
Plate: emissivity = 0.9
Covers: transmittance = 0.8, absorptivity = 0.1 = emissivity
Insulation: thickness 50 mm, K = 0.10 W/m K
Convection coefficient between the plate and cover, and the covers = 100 W/m² K; between the top cover and the air = 50 W/m² K
Assuming that the plate temperature = 400 K, derive an expression for the efficiency and heat collected.

Solution Referring to the previous example,

R_1 = thickness/KA_c = 0.05/0.10 × 5 = 0.10
$R_3 = 1/A_c(h_{c2} + h_{r2})$

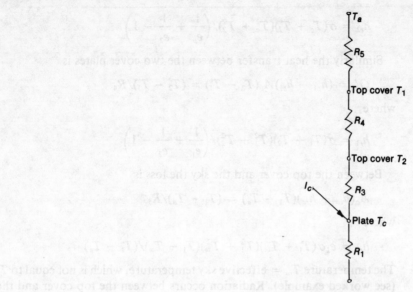

Figure 9.2

$$h_{r2} = \sigma(T_c + T_2)(T_c^2 + T_2^2)/\left(\frac{1}{e_p} + \frac{1}{e_2} - 1\right)$$

where

$$1/e_p + 1/e_2 - 1 = \frac{1}{0.9} + \frac{1}{0.1} - 1 = 10.11$$

$$R_4 = 1/A_c(h_{c1} + h_{r1})$$

where

$$h_{r1} = \sigma(T_1 + T_2)(T_1^2 + T_2^2)/\left(\frac{1}{e_1} + \frac{1}{e_2} - 1\right)$$

where

$$\frac{1}{e_1} + \frac{1}{e_2} - 1 = \frac{1}{0.1} + \frac{1}{0.1} - 1 = 19.00$$

$$R_5 = 1/A_c(h_{ca} + h_{ra})$$

where

$$h_{ra} = e_1\sigma(T_1 + T_\infty)(T_1^2 + T_\infty^2)(T_1 - T_\infty)/(T_1 - T_a)$$

Also

$$\frac{T_c - T_2}{R_3} = \frac{T_2 - T_1}{R_4} = \frac{T_1 - T_a}{R_5}$$

Now, $R_1 = 0.10$

$R_3 = \frac{1}{5}(100 + h_{r2})$

178 SOLVING PROBLEMS IN APPLIED THERMODYNAMICS AND ENERGY CONVERSION

$$R_4 = \tfrac{1}{5}(100 + h_{r1})$$
$$R_5 = \tfrac{1}{5}(50 + h_{ra})$$
$$h_{r2} = \sigma(T_c + T_2)(T_c^2 + T_2^2)/10.11$$
$$h_{r1} = \sigma(T_1 + T_2)(T_1^2 + T_2^2)/19.00$$
$$h_{ra} = 0.1\sigma(T_1 + 273)(T_1^2 + 273^2)$$
$$\times (T_1 - 273)/(T_1 - 283)$$

Hence for $T_c = 400$ K,

$$\frac{400 - T_2}{R_3} = \frac{T_2 - T_1}{R_4} = \frac{T_1 - 283}{R_5}$$

These can be solved for T_1 and T_2.

The overall coefficient U_c can then be obtained, and finally the efficiency

$$\eta_c = 0.8(0.9) - U_c(400 - 283)/800 = \mathbf{0.72 - 0.146\,U_c}$$

and heat collected

$$Q_u = 800(5)(0.8)(0.9) - U_c(5)(400 - 283)$$
$$= \mathbf{2880 - 585\,U_c\,W}$$

9.4 Solar cell

Briefly outline the principle of a solar (photovoltaic) cell.

At an intensity of 950 W/m² on the cell the short-circuit current density is 180 A/m² and the reverse saturation current density is 8×10^{-9} A/m². Temperature = 27 °C. If the output = 10 W, determine the surface area needed, and the conversion efficiency. Assume maximum power output conditions.

$e/K = 11\,600$ where e = electron charge, K = Boltzmann constant.

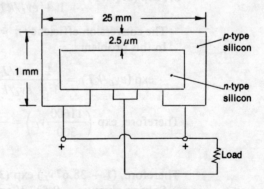

Figure 9.3

Solution The incident photons that are not reflected enter the thin layer of semiconducting material and are converted into heat or produce pairs of ions. Some of the ions are separated by the electric field of the p–n junction, reducing the field at the junction and producing a current flow.

The base material is often silicon, doped with phosphorus to produce the n-type and boron to produce the p-type semiconductors. Other materials being investigated include cadmium sulphide, cadmium tellurium, cadmium selenium.

Cells are simple, compact, have a high power/weight ratio, no moving parts, and long life but are very expensive.

The net current across the junction is

$$J = J_0 \exp\left[(ev_L/kT) - 1\right]$$

where J_0 = reverse saturation current density
K = Boltzmann constant = 1.55×10^{-4} ev/K
T = temperature (K)
e = charge of an electron.

If the photoelectric effect is J_s, then for any load on the cell, part is shunted through the internal resistance of the cell, or external load current density $J_L = J_s - J$.

The power output from the cell, area A, is

$$P = v_L J_L A = v_L A \left[J_s - J_0 \left(\exp\frac{ev_L}{kT} - 1\right)\right]$$

For maximum power $dP/dv_L = 0$, therefore

$$0 = AJ_s - AJ_0\left[\left(\exp\frac{ev_L}{kT} - 1\right) + v_L\frac{e}{kT}\left(\exp\frac{ev_L}{kT}\right)\right]$$

Therefore, $\exp\dfrac{ev_L}{kT}(J_0 + J_0 ev_L/kT) = J_0 + J_s$

or, $\exp(ev_L/kT) = \dfrac{1 + J_s/J_0}{1 + ev_L/kT}$

The conversion efficiency $\eta = P/P_{in}$.
In this problem

$$\exp(ev_L/kT) = \frac{1 + J_s/J_0}{1 + ev_L/kT}$$

Therefore, $\exp\left(\dfrac{11600}{300}v_L\right) = \dfrac{1 + 180/(8 \times 10^{-9})}{1 + \dfrac{11600}{300}v_L}$

Therefore, $(1 + 38.67 v_L)\exp(38.67 v_L) = 22.5 \times 10^9$.
Solving gives $v_L = 0.537$ V and

$$P_{mzx}/A = v_L\left[J_s - J_0\left(\exp\frac{ev_L}{kT} - 1\right)\right]$$
$$= 0.537[180 - 8 \times 10^{-9}(1.043 \times 10^9 - 1)]$$
$$= 92.18 \text{ W/m}^2$$

and cell area = 10/92.18 = **0.108 m²**.
The conversion efficiency = 92.18/950 = **0.097**.

9.5 Windmills

> Briefly outline how the power coefficient of a windmill varies with the speed ratio. Show that the ideal value of the maximum power coefficient is 16/27, and determine the efficiency at this condition.
> Estimate the ideal power generation (maximum) from a windmill with blades 50 m diameter at a wind speed of 5 m/s.

Solution The power coefficient C_p is defined as the power output/energy of the free stream (approach wind speed),

$$C_p = P \bigg/ \left[\left(\tfrac{1}{2}\rho v_\infty^3\right)\left(\frac{\pi D^2}{4}\right)\right]$$

Typical curves are shown in Fig. 9.4.

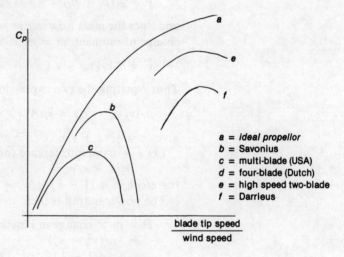

a = ideal propellor
b = Savonius
c = multi-blade (USA)
d = four-blade (Dutch)
e = high speed two-blade
f = Darrieus

Figure 9.4

The rotor is considered to act as an actuator disc, across which there is a pressure drop, and through which a mean velocity v can be considered.
Assuming incompressible flow

$$p_\infty + \tfrac{1}{2}\rho v_\infty^2 = p_1 + \tfrac{1}{2}\rho v_1^2$$
$$p_2 + \tfrac{1}{2}\rho v_2^2 = p_e + \tfrac{1}{2}\rho v_e^2$$

Therefore

$$p_1 - p_2 = \tfrac{1}{2}\rho(v_\infty^2 - v_e^2) - \tfrac{1}{2}\rho(v_2^2 - v_1^2) + (p_\infty - p_e)$$

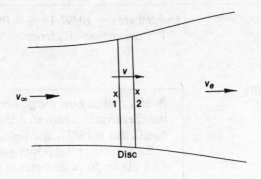

Figure 9.5

Assuming that $p_e = p_\infty$, and neglecting $v_2^2 - v_1^2$ gives

$$p_1 - p_2 = \tfrac{1}{2}\rho(v_\infty^2 - v_e^2)$$

The force on the disc

$$F = A(p_1 - p_2) = \tfrac{1}{2}\rho A(v_\infty^2 - v_e^2)$$

and since the mass flow rate $m = \rho A v$, $F = m(v_\infty - v_e)$, (force = rate of change of momentum = $m \times$ velocity change), therefore

$$F = \rho A v(v_\infty - v_e)$$

Thus equating the two expressions for the force

$$\rho A v(v_\infty - v_e) = \tfrac{1}{2}\rho A(v_\infty^2 - v_e^2)$$

gives $v = \tfrac{1}{2}(v_\infty + v_e)$.

Let λ = axial interference (or perturbation) factor
$= 1 - v/v_\infty$
therefore, $v = (1 - \lambda)v_\infty$, $v_e = 2v - v_\infty = (1 - 2\lambda)v_\infty$.

The power output is then

$$\begin{aligned}P &= m \times \text{change in kinetic energy} \\ &= \tfrac{1}{2}m(v_\infty^2 - v_e^2) \\ &= \tfrac{1}{2}\rho A(1 - \lambda)v_\infty \cdot v_\infty^2[1 - (1 - 2\lambda)^2] \\ &= 2\rho A v_\infty^3 \lambda(1 - \lambda)^2\end{aligned}$$

For a given area and wind velocity v_∞ the power will be a maximum when $dP/d\lambda = 0$, or $\dfrac{d}{d\lambda}(\lambda - 2\lambda^2 + \lambda^3) = 0$, therefore $1 - 4\lambda + 3\lambda^2 = 0$ giving $\lambda = \tfrac{1}{3}$, and at this condition

$$P_{max} = \tfrac{8}{27}\rho A v_\infty^3,\; C_p = P/\tfrac{1}{2}\rho A v_\infty^3 = \mathbf{16/27}$$
$$\eta = P/\tfrac{1}{2}mv_\infty^2 = P/\tfrac{1}{2}\rho A(1-\lambda)v_\infty^3 = \mathbf{24/27}$$

At $A = \pi(25)^2$, $v_\infty = 5$ m/s and an air density of 1.2 kg/m³, maximum power = $8(1.2) \times \pi(25)^2(5)^3/27 = 87267$ W or **87.3 kW**.

9.6 Geothermal energy

> Discuss the problems involved in the utilization of geothermal energy.
>
> Outline how the available energy fraction can be estimated in terms of the initial reservoir temperature T_1, economical temperature at which production ceases T_2.

Solution The geothermal energy is available in one of three forms, viz. dry steam fields (Lardarello, Geysers), much more abundant wet-steam fields (Wairakei), and low-temperature water fields (Kamchatka).

The steam from dry fields can be used directly in a steam turbine, but the pressure is low (up to 15 bar), and the temperature is low (up to 250 °C). Hence the efficiency is low. Also the steam rapidly becomes wet in passing through the turbines, and there can be problems with mineral salts.

The low-temperature water (50–80 °C) can be used directly for heating, or for power generation using a heat engine employing Freon as the working fluid.

There are also environmental problems in that land subsidence can occur as the steam/water is extracted. For example, in the Wairakei field some 70 million tonnes of water are extracted each year, and the field is gradually changing from a wet-steam to a dry-steam field.

The resource available is $C_v(T_1 - T_0)/m^3$, where C_v = specific heat = 2500 kJ/m^3 (average).

The ideal fraction recoverable is $(T_1 - T_2)/(T_1 - T_0)$, which becomes, allowing for Carnot limitations:

$$\frac{T_1 - T_2}{T_1 - T_0} \times \left(1 - \frac{T_0}{T_m}\right)$$

where $T_m = (T_1 - T_2)/\ln(T_1/T_2)$.

9.7 Refuse incineration

> The analysis of a domestic refuse is, by mass: 11 % cinder, 4 % putrescible matter, 40 % paper, 27 % metal, 15 % plastics, 3 % rags.
>
> The CV of the combustibles are, in MJ/kg, cinder 27, paper 17, plastics 19, rags 13.
>
> The refuse collected in a population of 6000 is 6 kg/week per capita.
>
> The weekly refuse is burnt in a boiler to raise steam, from feedwater at 20 °C, at 20 bar, 250 °C.
>
> Determine the mass of steam generated, if the boiler efficiency is 0.6.
>
> Outline the problems involved in the incineration of refuse, and the benefits of this system.

Solution The CV of the fuel/kg = $0.11(27) + 0.4(17) + 0.15(19) + 0.03(13) = 13.0$ MJ.

Therefore, heat absorbed in steam = $0.6 \times 6 \times 6000 \times 13 = 280\,800$ MJ = $m(2904 - 83.9)$.

Therefore, steam generated = **99.6 kg/week/capita**.

The problems involved are summarized as follows:

(a) the material is usually damp;
(b) the contents are highly variable;
(c) incombustible components need to be removed before combustion;
(d) incinerators are expensive to build, maintain and operate, and tend to be rather inefficient;
(e) there are problems of corrosion and fouling in the combustion chamber;
(f) noxious products may be present in flue gases.

The benefits include:

(a) no use of land for tipping purposes,
(b) produces sterile residue which can be used as hardcore or other filler material,
(c) encourages the recovery of metals, glass, etc.

9.8 Fuel-consumption rates

The annual fuel consumption at a certain time is 3×10^6 tonnes, and the fuel reserves are estimated at 8×10^{12} tonnes.

Assuming that the annual increase remains constant, and gives a consumption doubling time of 20 years, how long would the reserves last (assuming that they can be wholly extracted)?

The fuel in question is coal, and the figures quoted are on a global basis. Comment on the value obtained for the life of the coal.

Solution Let S_0 = initial consumption rate, $t = 0$
r = annual rate of increase
S_n = consumption in nth year, t_n

Then considering an exponential growth rate $S_n = S_0 \exp(rt_n)$. At time $t = t_1$, $S_1 = S_0 \exp(rt_1)$ and at time t_2, $S_2 = S_0 \exp(rt_2)$; hence $S_2/S_1 = \exp r(t_2 - t_1)$. The doubling time, i.e. when $S_2 = S_1$, is therefore given by

$$2 = \exp r(t_2 - t_1) \quad \text{and} \quad t_2 - t_1 = \ln(2/r) = 0.693/r$$

The amount of fuel consumed, E, can be determined by integration

$$E = \int S \, dt$$

therefore $E_2 - E_1 = \int_{t_1}^{t_2} S \, dt = \int_{t_1}^{t_2} S_0 \exp(rt) \, dt$

$$= \frac{1}{r} S_0 [\exp(rt)]_{t_1}^{t_2}$$

$$= \frac{S_0}{r} \exp r(t_2 - t_1)$$

$$= \frac{S_0}{r} \exp (rt_1)[\exp r(t_2 - t_1) - 1]$$

Hence, if $t_2 - t_1$ is the doubling time,

$$E_2 - E_1 = \frac{S_0}{r} \exp (rt_1)(2 - 1) = \frac{S_0}{r} \exp (rt_1)$$

Considering the consumption for *all* time prior to the doubling period

$$E_1 = \int_{-\infty}^{t_1} S \, dt = \frac{S_0}{r} \exp (rt_1)$$

Thus the consumption in one doubling period is equal to the consumption during all the period prior to this doubling period. This is shown graphically in Fig. 9.6.

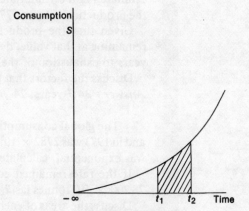

Figure 9.6

Area under the curve

$$= \int S \, dt = E$$

If $t_1 - t_2$ = doubling time the two areas shown are equal.

In this problem the doubling time = 20 and $S_0 = 3 \times 10^6$. Therefore, $20 = 0.693/r$ and $r = 0.0346$ or **3.46 %**. The reserves would then be theoretically exhausted when

$$8 \times 10^{12} = \frac{S_0}{r} \exp (rt_2) - \frac{S_0}{r} \exp (rt_1)$$

$$= \frac{3 \times 10^6}{0.0346} [\exp 0.0346t - \exp 0]$$

Therefore $9.227 \times 10^4 = \exp 0.0346t - 1$

Therefore $t = $ **330 years**.

The reserves of coal are not distributed equally throughout the world, and the usage of coal is not likely to remain constant. The rate of increase of coal consumption has decreased dramatically from the beginning of the twentieth century, as can be seen from the proportion of the global fossil fuel supply taken by coal. This has fallen from about 95 % at the beginning of the century to about a constant figure of 30 % at present. The use of coal may increase, however, as reserves of petroleum and natural gas become depleted, requiring their substitution by processing from coal stocks.

There are also questions of partial recovery and uneconomic seams, and the market forces involved as any fuel reserve becomes depleted.

Problems

1 The offshore reserves of oil in the North Sea (UK) sector were estimated at a certain time to be 2.4×10^9 tonnes, and at that time the production rate was 77×10^6 tonnes.

Given that the production rate increased at 2 % per annum, remaining at that value, determine the doubling time and number of years to exhaustion of the reserves.

Discuss the factors that prevent this simple scenario taking place.
Answer 34.7 years; 24 years

2 The global consumption of oil in 1970 was 2287×10^6 tonnes, and in 1973 was 2782×10^6 tonnes. Assuming that the rate of increase was exponential, calculate the rate of increase and doubling time.

If the rate remained constant how long would the reserves of $78\,200 \times 10^6$ tonnes last?

Discuss the areas of energy utilization in which a viable alternative to oil fuels would be problematical. Outline the alternative energy sources that have been suggested.
Answer 0.065; 16 years; 18 years

3 Conservation measures in a country are expected to reduce the electricity consumption. The current consumption is 1.5×10^{12} kWh, with a doubling time of 10 years. The measures are such that the doubling time is doubled every doubling period.

Compare the demand after 70 years to the demand if no conservation was employed, and the total amount of electricity used in each case.
Answer 12 and 192×10^{12} kWh; 990 and 2770×10^{12} kWh

4 The production rate of coal follows a normal distribution curve, given by

$$P = 20 \times 10^6 \exp[-(t/200)^2] \text{ tonnes/year}$$

where t = time (years), measured from the time of peak production.

Plot a graph of production rate against time, over a period of -350 to $+350$ years: and also a curve showing the cumulative consumption of coal.

If the reserves of coal were 6×10^9 tonnes, estimate the time when they would be exhausted.

Briefly outline the factors in world coal production and reserves situation that would affect this simplified analysis.
Answer $+190$ years

5 A fuel reserve is estimated to last 420 years at the current consumption rate. How long would it last if the consumption increased at a rate of 2.6 % per year? Determine the doubling time.

10 years hence the reserves, at that time, are doubled by new discoveries. The increase rate remains the same at 2.6 % per annum. Estimate the life of the new reserves.
Answer 95, 26.7, 120 years

6 A solar collector is used to generate electric power. The incident radiation is 600 W/m². Overall coefficient of heat transfer = 4.0 W/m² K. Collector efficiency = 0.70. The mass flow rate of water through the collector is 0.008 kg/m² s. The effective transmittivity $\tau\alpha$ = 0.8. Ambient temperature = 20 °C. Collector area = 50 000 m².

Assuming that the efficiency of the heat engine is one-third that of a Carnot engine working between the ambient and collector outlet temperatures estimate the power generated. The collector efficiency $= (\tau\alpha) - U(T_c - T_a)$ where T_c = mean water temperature.

Outline an alternative method of using solar energy to generate electrical power.
Answer 480 kW

7 A test on a solar collector gave the results shown in Table 9.1.

Table 9.1

Run	I_c (W/m²)	T_a (°C)	T_w (°C) in	T_w (°C) out	Water flow m (kg/s)
1	570	27	54	58	0.0295
2	740	30	45	51	0.0287
3	860	33	43	51	0.0287
4	920	34	65	72	0.0310
5	940	36	65	72	0.0310
6	950	37	86	92	0.0326
7	930	38	96	101	0.0333
8	855	39	97	101	0.0333
9	730	39	98	101	0.0340

In the table, I_c = incident radiation, T_a = ambient temperature. The efficiency $\eta = (\tau\alpha) - U(T_c - T_a)/I_c$, where T_c = mean water temperature. The collector area = 1.6 m^2.

Plot the results and hence determine a mean value for the effective transmittivity $\tau\alpha$, and the overall coefficient U.

Answer 0.8, 6 W/m² K

8 A flat-plate collector can be considered, in a simple model, as a sheet and black collector surface. Show that in the case of n sheets

$$(n + 1)(I_c + Q_u) = \sigma(T_o^4 - T_a^4)$$

where I_c = incident solar energy/m²
Q_u = energy removed/m²
T_o = collector surface temperature
T_a = ambient temperature.

Also show that $T_o^4 = T_a^4 + (n + 1)(I_c/\sigma)(1 - \eta)$, where η = collector efficiency = Q_u/I_c.

It may be assumed that the cover sheet does not absorb any solar energy.

Determine the maximum temperature that could be attained when $T_a = 0\,°\text{C}$, $I_c = 190 \text{ W/m}^2$, $\eta = 0.50$. Consider (a) one, (b) two cover sheets.

What temperature could be reached if a focusing collector (concentrator) was used, of area ratio 100:1?

Answer 34, 48; 367 °C

9 A solar energy cooling system uses solar collectors, at a temperature T_2, to drive a heat engine. The heat is rejected at temperature T_1 to ambient air (which is at a temperature T_a). The engine drives a refrigerator which extracts heat (at temperature T_1), rejecting heat at temperature T_2, to ambient air.

Assuming that the engine and refrigerator operate on a Carnot cycle, show that

$$Q_1/Q_2 = \left(\frac{T_1}{T_a - T_1}\right)\left(\frac{T_2 - T_a}{T_2}\right)$$

where Q = heat corresponding to a temperature T.

Given that $T_a = 30\,°\text{C}$, $T_2 = 100\,°\text{C}$, $T_1 = 10\,°\text{C}$ and the collector power density = 600 W/m², estimate the collector area required to remove 100 kW of heat.

Answer 0.063 m²

10 A concentration solar collector plant is to be used for the generation of electric power, and the load required/day is 2400 kWh. The incident solar radiation is 600 W/m² and is available for 8 hours/day. The collector efficiency = 0.5, and the efficiency of conversion from thermal to electrical energy is 0.3.

Estimate the collector area required.

The cost of electricity, purchased directly, is 5 p/kWh. Determine the cost of collector/m^2 if the solar generated power is at the same cost as the direct purchase power.

Briefly outline the factors involved in this type of solar-energy conversion that affect the economics and viability.

Answer 3330 m^2; 3.6 p/m^2

11 Discuss the use of biomass as an energy source, commenting on the possible applications and problems involved.

The production of biomass in a certain region is 30 tonnes/acre (7.4 kg/m^2). The calorific value of the biomass is 20 MJ/kg. The conversion efficiency (biomass − electric power) is 0.30. To get an output of 10^6 kWh estimate the area of land required.

Answer 81 080 m^2

12 Briefly outline the method of using tidal energy for electric power generation. Discuss the advantages of using a two-pool system.

A tidal power plant uses two pools: the upper pool, of area A_1, is filled at high tide, and the lower pool, of area A_2, is emptied at low tide.

Determine the energy available if the upper pool is emptied into the lower pool, with the gates connecting both pools to the ocean closed.

The scheme is to be compared with a single-pool scheme, of area $A = A_1 + A_2$.

Determine the ratio of the energy available from each scheme; and the value of this ratio when $A_1 = A_2 = \frac{1}{2}A$.

13 An ocean thermal energy conversion plant (OTEC) proposed by Claude in 1929 consisted of a Rankine cycle plant, using sea water at reduced pressure. The boiler temperature = 25 °C (at 0.0317 bar), and the condensor temperature = 15 °C (at 0.0170 bar). The steam is dry saturated at the turbine inlet.

Assuming a temperature drop of 2 °C in the condensor, estimate the warm and cold water flow rates required for an output of 100 MW.

Outline how the scheme could be imposed.

Answer 28.9; 347 000 kg/s

14 The wind velocity distribution over a period of time T varies in four ways:

(a) $v = 2\bar{v}, 0 < t < \frac{1}{2}T$: $v = 0, \frac{1}{2}T < t < T$
(b) $v = 3\bar{v}, 0 < t < \frac{1}{3}T$: $v = 0, \frac{1}{3}T < t < T$
(c) $v = \bar{v}(1 + \sin 2\pi t/T) : 0 < t < T$
(d) $v = 2\bar{v}(t/T)$ $: 0 < t < T$

where $\bar{v}$ = average velocity.

Determine for each distribution the ratio of the energy available to the energy available from a steady wind at a velocity $\bar{v}$.

Outline the difficulties involved in the use of a windmill (aerogenerator) to extract energy from the wind.

Answer 1.0 in each case

10

Waste-heat recovery; total energy: combined heat and power; energy economics

The efficient use of energy can be considerably improved by the use of combined heat and power generation (CHP), the use of waste heat recovery, insulation, careful control and audits (monitoring).

Many prime movers reject heat at substantial temperature levels, which can be utilized for heat recovery, or lower-grade heat utilization. The waste heat produced in electricity generation could be used for space and water (district) heating for example.

Combined cycles (co-generation) normally produce power and heat, but can produce power only, and are normally associated with steam/gas turbines. A gas turbine is used to generate power; the exhaust heat is recovered in a waste-heat boiler and the steam raised is then passed through a steam turbine to a condensor or process load.

Low-grade waste heat can also be used for process and heating applications, as for example, in horticulture, fish-farming, space heating.

Prime movers Various types of steam turbine are available:

(a) back-pressure: cheap, compact, but inflexible;
(b) condensing: expensive, bulky;
(c) pass-out condensing: commonest type, flexible.

Other prime movers include:

(a) diesel engines: burning diesel oil, fuel oil;
(b) gas turbines: burning natural gas, waste gases, oil.

Energy costs The total cost for a plant is made up of capital, fuel and operational costs. The capital cost is constant and must be paid whether the plant operates or not, and therefore the cost per kilowatt-hour generated decreases as the output increases. It includes the cost of the land for the site, construction cost, taxes, insurance and interest on the investment. The operational cost includes wages, maintenance and fuel costs.

In the consideration of alternative schemes the saving in energy costs is balanced against the capital cost, changes in maintenance and labour charges, and interest costs. The simple payback period, without a detailed economic analysis, can be taken as the initial capital cost/annual saving in energy costs.

10.1 Pass-out condensing plant

A pass-out condensing steam turbine is used to generate electric power and supply process heat. The plant is shown in Fig. 10.1 and the data in Table 10.1.

Table 10.1

Boiler: coal fired
  CV coal = 25 mJ/kg
  efficiency = 0.8
Turbine: isentropic efficiency
  (both sections = 0.70)

Steam conditions:	Point	Condition
	1	30 bar, 500 °C
	2	6 bar
	3	6 bar, saturated liquid
	4	0.20 bar

Electric power output = 10 MW
Process heat = 30 MW

This scheme is intended to replace an existing plant, in which the electricity is purchased from the grid at a cost of 2.5 p/kWh, and the process heat obtained from a boiler of thermal efficiency 0.65.

Assuming that the cost of coal = £40/tonne and that the plants are operated for 6000 h during a year, compare the two plants for annual fuel cost and overall thermal efficiency.

Solution Let the pass-out flow at point 2 be km kg/s. Considering the steam turbine,

$$h_1 = 3456 \text{ kJ/kg}, \quad s_1 = 7.233 \text{ kJ/kg K}$$

By calculation, or using the Molier chart, for isentropic expansion $h_{2s} = 2986$, therefore $\Delta h_s = 470$ and $\Delta h = 0.7 \times 470 = 329$ and $h_2 = 3456 - 329 = 3127$ kJ/kg, $s_2 = 7.48$ kJ/kg.

For the expansion from points 2 to 4: $h_{4s} = 2467$ kJ/kg, $\Delta h_s = 3127 - 2467 = 660$ and $\Delta h = 0.7 \times 660 = 462$, therefore $h_4 = 2665$ kJ/kg.

The electrical power output is

$$E = m(h_1 - h_2) + (1 - k)m(h_2 - h_4) \text{ kW}$$

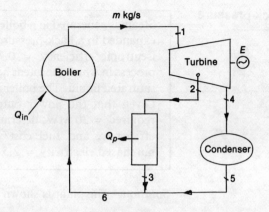

Figure 10.1

Therefore, $10\,000 = 329m + 462(1 - k)m = (791 - 462k)m$.

At point 3, $h_f = 670$ kJ/kg and the process heat = 14 MW, therefore $30\,000 = km(h_2 - h_3) = 2457km$.

Hence $k = 0.617$, $m = $ **19.79 kg/s**.

Considering the boiler,

$$\text{efficiency} = 0.8 = m(h_1 - h_6)/m_f \cdot CV$$

where m_f = coal firing rate (kg/s). The enthalpy of the boiler feed, h_6, can be determined by assuming adiabatic mixing of the process return water (h_3) and condensate (h_5)

$$kmh_3 + (1 - k)mh_5 = mh_6$$

therefore, $h_6 = 0.617(670) + 0.383(251) = 510$ kJ/kg.

Hence, $m_f = 19.79(3456 - 400)/(0.8 \times 25\,000) = 3.02$ kg/s.

Overall thermal efficiency of plant = $(E + Q_p)/Q_{in} = (10 + 30)/(3.02 \times 25) = $ **0.53**.

Annual fuel cost = £3.02 × 3600 × 6000 × 10^{-3} × 40 = **£2 609 300**.

In the existing scheme, the process heat is supplied by steam generated at 6 bar in a boiler. Assuming that the steam entering the process plant is at the same condition ($h = h_2 = 3127$ kJ/kg) the flow rate of steam will be $30\,000/(3127 - 670) = 12.21$ kg/s, and fuel input to the boiler = $30\,000/(0.65 \times 25\,000) = 1.85$ kg/s, therefore fuel cost/year = £1.85 × 3600 × 6000 × 10^{-3} × 40 = £1 598 400.

The annual cost of the purchased electricity is $10\,000 \times 6000 \times 2.5$ p = £1 500 000 and the total cost = **£3 098 400**.

The overall thermal efficiency can be estimated by assuming that the electricity is generated and distributed with a thermal efficiency of 0.30. The total heat input is then $10/0.3 + 30/0.65 = 79.49$ MW, and overall efficiency = $(10 + 30)/79.49 = $ **0.50**.

The replacement scheme saves £489 100/year on energy costs.

10.2 Back-pressure turbine

Steam is generated in a boiler (efficiency 0.8) at 30 bar, 500 °C. It is expanded in a back-pressure turbine to a pressure of 6 bar (turbine isentropic efficiency = 0.70). The steam then passes through a process in which the latent heat is extracted, returning to the boiler as saturated liquid. The boiler is coal fired (CV for coal = 25 MJ/kg). Given that the power output = 10 MW, and the process heat required = 30 MW, determine the steam flow rate, overall thermal efficiency, and fuel cost/year. Cost of coal = £40/tonne: of purchased electricity = 2.5 p/kWh.

Solution The plant is shown in Fig. 10.2.

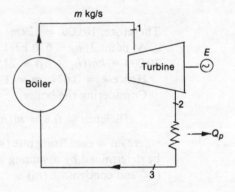

Figure 10.2

Referring to the previous example, 10.1, h_1 = 3456 kJ/kg
h_2 = 3127 kJ/kg
h_3 = 670 kJ/kg

In this plant the steam expands through the turbine to a back pressure of 6 bar.

Power output $E = m(h_1 - h_2)$
$= 329m$ kW

Process heat $Q_p = m(h_2 - h_3)$
$= 2457m$ kW

Hence, the two demands $E = 10\,000$, $Q_p = 30\,000$ cannot be satisfied together. If the steam flow rate is made to suit the power demand, $m = 10\,000/329 = $ **30.4 kg/s** giving $Q_p = $ **74 690 kW**, which is in excess of the demanded 30 000 kW.

If the steam flow rate suits the process heat demand $m = 30\,000/2457 = $ **12.21 kg/s**, giving a power output of 4017 kW and there will be a shortfall

194 SOLVING PROBLEMS IN APPLIED THERMODYNAMICS AND ENERGY CONVERSION

of 5983 kW. Considering this alternative, coal firing rate = 30 000/(0.8 × 25 000) = 1.50 kg/s, therefore fuel cost/year = 1.50 = 3600 × 6000 × 10^{-3} × 40 = **£1 296 000**.

The cost of purchased electricity = £5983 × 6000 × 2.5/100 = £897 450, giving a total cost of **£2 193 450**.

Overall thermal efficiency = 40/(1.5 × 25) + (5.983/0.3) = **0.70**.

10.3 Co-generation with variable load

A plant requires 20 MW electrical power and a heat load which varies per year (6000 h) as shown in Table 10.2.

Table 10.2

Time (h)	1000	2000	3000
Load (MW)	10	20	50

The existing plant comprises a boiler (efficiency 0.75), coal fired (*CV* 25 MJ/kg) to supply the heat load, and electricity is purchased at 2.5 p/kWh. Assuming that the heat is supplied as dry steam at 10 bar, and the return from the process is saturated liquid at 5 bar, calculate the energy cost/year. Cost of coal = £40/tonne.

It is proposed to replace the plant by a pass-out condensing turbine, exhausting at 0.10 bar, and a boiler generating steam at 30 bar, 295 °C. Assuming a straight condition line, estimate the annual fuel cost of the plant.

Comment on the values obtained for the fuel costs.

Would a back-pressure turbine be more suitable?

Solution At any heat load Q, the steam flow rate m is given by $m(2778 - 640) = Q$ kW. The corresponding coal rate $m_f = Q/(0.75 \times 25\,000)$ kg/s, and the fuel used/year is

$$\frac{3600}{0.75 \times 25\,000}(10\,000 \times 1000 + 20\,000 \times 2000 + 50\,000 \times 3000) \text{ kg}$$

therefore

$$\text{fuel cost/year} = \frac{3600 \times 200 \times 10^6}{0.75 \times 25\,000} \times 10^{-3} \times £40 = £1\,536\,000$$

Cost of purchased electricity/year = £20 000 × 6000 × 2.5/100 = £3 000 000, therefore total cost = **£4 536 000/year**.

In the new scheme the steam flow rate passed out at 10 bar will vary with the heat demand Q.

The steam conditions at point 2 are dry saturated, and $h_2 = 2778$ kJ/kg. The enthalpy at 3 is 640 kJ/kg. At the turbine inlet $h_1 = 2980$ kJ/kg.

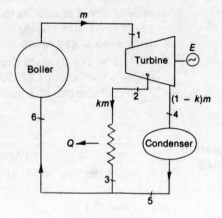

Figure 10.3

From the Molier chart a straight condition line gives $h_4 = 2180$ kJ/kg.
Power output $E = 20\,000 = m(h_1 - h_2) + (1 - k)m(h_2 - h_4)$ kW
$$= 800m - 598km$$
Heat load $Q = km(h_2 - h_3)$ kW $= 2138\,km$.

The coal rate m_f is then determined from

$$\frac{m(h_1 - h_6)}{0.75 \times 25\,000} \text{ kg/s}$$

where for adiabatic mixing

$$kmh_3 + (1 - k)mh_5 = mh_6$$

therefore, $h_6 = 640k + 192(1 - k)$ kJ/kg.

Q (kW)	10 000	20 000	50 000
m (kg/s)	29.23	32.26	42.52
h_6 (kJ/kg)	264	322	438
m_f (kg/s)	4.23	4.57	5.76
time (h)	1000	2000	3000
m_f (tonnes)	15 228	32 904	62 208
k	0.16	0.29	0.55

Total coal consumed/year $= 15\,228 + 32\,904 + 62\,208 = 110\,340$ and cost/year $= 110\,340 \times 40 =$ **£4 413 600**.

The replacement plant thus saves £122 400/year on fuel costs. Whether this would be economic or not depends on the capital cost of the turbo-alternator, condenser and boiler, and differences in operating costs. Since the saving in fuel costs is 2.7 % per year it would probably be uneconomic to replace the plant.

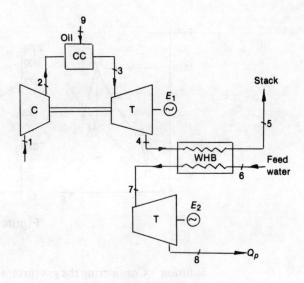

Figure 10.4

10.4 Co-generation with waste-heat boiler

A co-generation plant is shown in Fig. 10.4. The conditions at the points are given in Table 10.3.

Table 10.3

		m (kg/s)	T (°C)	p (bar)
1	air	60	20	1.0
2	air	60	?	10
3	gas	61	1000	10
4	gas	61	?	1
5	gas	61	150	1
6	water	?	100	
7	steam	?	400	40
8	steam	?	?	3
9	oil	1		

The gas-turbine compressor and turbine isentropic efficiencies are 0.80 and 0.85 respectively, and that of the steam turbine 0.70.

Determine the gas-turbine cycle efficiency, the steam flow generated in the waste heat boiler, the power output from the gas and steam turbines, the process heat in the steam at the turbine exhaust, and overall thermal efficiency of the plant.

Outline how the heat exchange area required in the waste heat boiler can be estimated, and explain the term 'pinch point'.

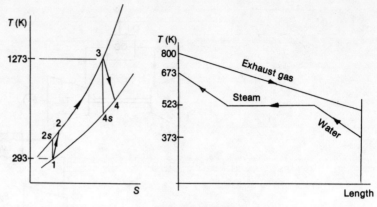

Figure 10.5

Solution Considering the gas turbine cycle (Fig. 10.5), $T_{2s} = 293(10)^{0.286} = 566$ K, $T_2 = 293 + (566 - 293)/0.8 = 634$ K, $T_{4s} = 1273/(10)^{0.25} = 716$ K, $T_4 = 1273 - 0.85(1273 - 716) = 800$ K.

$$\text{Cycle efficiency} = \frac{1.15(1273 - 800) - 1.005(634 - 293)}{1.15(1273) - 1.005(634)} = \mathbf{0.24}.$$

The air-flow rate = 60 kg/s, and fuel rate = 1 kg/s. Allowing for the differences in the flow rates,

$$\text{heat input} = 61(1.15)(1273) - 60(1.005)(634) = 51\,070 \text{ kW}$$
$$\text{net work (power output } E_1) = 61(1.15)(473) - 60(1.005)(341)$$
$$= 12\,620 \text{ kW}$$

and

$$\text{thermal efficiency} = 12\,620/51\,070 = 0.247$$

A heat balance on the waste heat boiler gives

$$m_4(h_4 - h_5) = m_7(h_7 - h_6)$$

therefore $61(1.15)(800 - 423) = m_7(3214 - 419)$, and the steam rate generated = **9.46 kg/s**.

The steam turbine expands the steam to a back-pressure of 3 bar. Using the Molier chart, or by calculation, $h_8 = 2808$ kJ/kg (slightly superheated).

The power output, $E_2 = 9.46(3214 - 2808) = \mathbf{3840 \text{ kW}}$

process heat, $Q_p = 9.46(2808) = \mathbf{26\,560 \text{ kW}}$

Overall thermal efficiency = $(E_1 + E_2 + Q_p)/51\,070 = \mathbf{0.84}$.

The waste heat boiler temperature distribution is shown in Fig. 10.5. The heat-transfer rate between the gases and the water is given by $Q = UA\,\Delta T_m$, where

A = exchange area U = overall coefficient ΔT_m = ln (mean temperature difference)

Thus for the three sections:

Water: $\Delta T_m = \dfrac{(T_a - 523) - 50}{\ln\,[(T_a - 523)/50]}$

Evaporation: $\Delta T_m = \dfrac{(T_b - 523) - (T_a - 523)}{\ln\,[(T_b - 523)/(T_a - 523)]}$

Superheater: $\Delta T_m = \dfrac{(T_b - 523) - 127}{\ln\,[(T_b - 523)/127]}$

The temperatures T_a, T_b depend on the slope of the gas temperature line (assumed linear).

The pinch point is the minimum temperature difference between the two fluids, $T_a - 523$. It is normally 20–30 °C, so that the heat transfer rate from the gas to the water does not become too small.

10.5 Diesel generator

A diesel engine is to be used for power generation and the heat in the cooling water and exhaust gas used for heating purposes. The power load is 10 MW.

Three types of engine are available (Table 10.4).

Table 10.4

	Capacity (MW)	Cost £/kW	Input (MW) I
A	7.5	150	$4 + 1.2L + 0.09L^2$
B	4	170	$2.7 + 1.2L + 0.15L^2$
C	2	180	$2.0 + 1.3L + 0.15L^2$

L = load (MW).

The fuel cost = 0.2 p/MJ, total annual cost = fuel cost + 0.15 × capital cost.

Assuming the heat in the cooling water and exhaust gas is 0.5 × heat input, and no standby sets are required, compare the annual cost and heat available from the different sets. Consider a 6000 h/year.

If the cost of purchased electricity is 2.5 p/kWh, and of heat 0.2 p/MJ, estimate the simple pay back time.

Comment on the values obtained.

Solution The power load of 10 MW will require 2 A sets, or 3 B sets, or 5 C sets.

The heat available = $0.5I$ MW, and the fuel cost is

$I \times 3600 \times 6000 \times 0.2/100 = £43\,200 I$ per year

The working is shown in Table 10.5.

Table 10.5

Item	$2 \times A$	$3 \times B$	$5 \times C$
Load/engine (MW)	5	3.33	2
I (MW)/engine	12.25	8.37	5.2
I (MW)	24.5	25.1	26.0
Heat Q (MW)	12.25	12.55	13.0
Fuel cost (£/year)	1 058 400	1 084 320	1 123 200
Capital cost (£) C_0	2 250 000	2 040 000	1 800 000
Total cost (£/year)	**1 395 900**	**1 390 320**	**1 393 200**
Purchased cost (£/year)	\multicolumn{3}{c}{$(10\,000 \times 6000 \times 2.5/100) +$}
	\multicolumn{3}{c}{$(Q \times 3600 \times 6000 \times 0.2/100) =$}
	2 029 200	2 042 160	2 061 600
Annual saving (£/year) S	633 300	651 840	668 400
Payback (years) = C_0/S	3.55	3.13	2.69

The heat available in the cooling water and exhaust gas is of the same order for all three alternatives, and the payback time is attractive for all three. However, a standby set would often be required, and this could move the selection more towards the use of the 2 MW sets. An additional set would cost: A £1 125 000, B £680 000, C £360 000. The maintenance and operating costs would probably be greater for the five smaller sets.

10.6 Co-generation with feed heating

A factory requires 5 MW electric power and 6 MW heat from the condensation of process steam, which leaves the process plant as saturated liquid at $5\frac{1}{2}$ bar.

The requirements are met by generating steam in a boiler (thermal efficiency 0.8) at 35 bar, 350 °C for expansion in an H.P. turbine to $5\frac{1}{2}$ bar. Process and feedwater heat is then extracted, and the remainder expands in the L.P. turbine to 0.07 bar, and is then condensed.

The extraction pump raises the condensate pressure to 1 bar, and it then enters the hot well at 1 bar, 40 °C. In the hot well it is mixed with the drain water from the feed heater, and make-up water supplied at 20 °C. The feedwater is then heated to 150 °C and pressurized to 35 bar.

The isentropic efficiency of the turbine is H.P. 0.85, L.P. 0.80. Boiler: coal fired, CV 25 MJ/kg, cost £40/tonne.

Determine the boiler evaporation rate, the flow rate to the feed heater, the heat input to the boiler, the coal firing rate, fuel cost/kWh power output, and overall thermal efficiency.

The heat removed from the condenser is used to provide space heating. The cooling water enters the condenser at 50 °C and is required to leave at 70 °C. Determine the water flow rate required.

Compare the annual fuel (energy) cost of the plant with a plant in which the process steam is generated in a boiler and the electricity purchased at a cost of 2.5 p/kWh. Assume a 6000 h year.

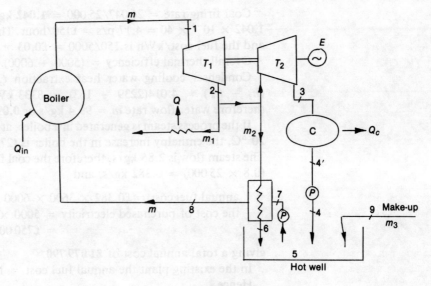

Figure 10.6

Solution   The plant layout is shown in Fig. 10.6.

The enthalpies at the various points in the cycle are obtained from steam tables and the Molier chart.

$h_1 = 3106, h_2 = 2760, h_3 = 2239$ kJ/kg
$h_4 = 168, h_6 = 656, h_8 = 628$ kJ/kg

The power output $E = 5000$ kW $= m(h_1 - h_2)$
$$+ (m - m_1 - m_2)(h_2 - h_3)$$
$$= 346m + 521(m - m_1 - m_2)$$

A heat balance for the feedwater heater gives

$$m(h_2 - h_6) = m(h_8 - h_7)$$

therefore, $2104 m_2 = (628 - h_7)m$.

A heat balance on the hot well gives

$$mh_5 = m_2 h_6 + (m - m_1 - m_2)h_4 + m_3 h_9$$

Now m_3 = make-up flow = m_1, $h_9 = 84$ kJ/kg; therefore

$$mh_5 = 656 m_2 + 168(m - m_1 - m_2) + 84 m_1$$

Solving the three equations gives the flow rates m, m_2 and h_5, taking $h_7 = h_5$, and using the condition that

$$Q = 6000 = m_1 h_{fg2} = 2097 m_1$$

therefore $m_1 = 2.86$ kg/s.

$m = \mathbf{8.409}$ **kg/s** $m_2 = \mathbf{1.535}$ **kg/s**   and $h_5 = \mathbf{244}$ **kJ/kg**

The heat input to the boiler

$$Q_{in} = m(h_1 - h_8)/0.8 = 8409(3106 - 628)/0.8 = \mathbf{26\,047\ kW}$$

Coal firing rate = 26 047/25 000 = **1.042 kg/s**; therefore fuel cost = $1.042 \times 10^{-3} \times 40 = 4.17$ p/s = **£150/hour**. The power output is 5000 W and the fuel cost/kWh is 150/5000 = £0.03 or **3 p/kWh**.

Overall thermal efficiency = $(5000 + 6000)/26\,047 = $ **0.42**.

Condenser cooling water heat extraction $Q_c = (m - m_1 - m_2)/(h_3 - h'_4) = 4.014(2239 - 163) = 8333$ kW $= m(4.19)(70 - 50)$, therefore water flow rate $m = 99.4$ kg/s or **0.099 m³/s**.

If the process steam is generated in a boiler, at $5\frac{1}{2}$ bar using feedwater at 20 °C, the enthalpy increase in the boiler is $(2753 - 84) = 2669$ kJ/kg. The steam flow is 2.86 kg/s, therefore the coal firing rate is $2.86 \times 2669/(0.8 \times 25\,000) = 0.382$ kg/s, and

$$\text{annual fuel cost} = £0.382 \times 3600 \times 6000 \times 10^{-3} \times 40 = £329\,790$$
$$\text{the cost of purchased electricity} = 5000 \times 6000 \times 2.5/100$$
$$= £750\,000/\text{year}$$

giving a total annual cost of **£1 079 790**.

In the existing plant the annual fuel cost = $150 \times 6000 = $ **£900 000**. Hence

$$\text{annual saving} = £179\,790 \quad \text{or} \quad (179\,790/900\,000) \times 100$$
$$= 20\,\%$$

10.7 Use of heat exchanger and heat pump

A launderette discharges 0.3 kg/s of water at 40 °C from the washing machines. The cold water supply, at 20 °C, is contained in a tank of capacity 5 tonnes.

Analyse the heating requirement in the following situations: the inlet temperature to the machines = 70 °C:

(a) direct heating of the cold water;
(b) use of a heat exchanger to extract heat from the discharge water;
(c) use of a heat exchanger and heat pump, $COP = 2$, and discharging from the evaporator at 10 °C.

The heat exchanger effectiveness = 0.7.

Solution The plant is shown in Fig. 10.7.

(a) $Q = mC_p(T - T_i)$, where $T = 70$, $T_i = 20$ °C; therefore

$$Q = 0.3(4.2)(50) = \mathbf{63\ kW}.$$

(b) The heat exchanger effectiveness $e = (T_d - T_e)/(T_d - T_i)$; therefore $T_e = T_d - e(T_d - T_i)$. A heat balance gives

$$mC_p(T_m - T_i) = mC_p(T_d - T_e)$$

therefore

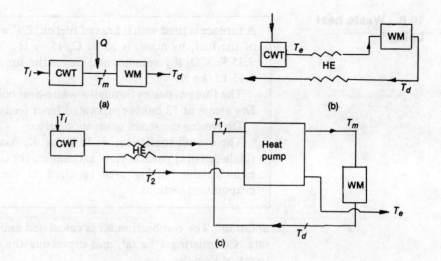

Figure 10.7

$$T_m = T_i + T_d - T_e = T_i + e(T_d - T_i)$$

Hence

$$T_m = 20 + 0.7(40 - 20) = 34 \,°C$$

The heat input is now reduced to

$$Q = 0.3(4.2)(70 - 34) = \mathbf{45.4 \text{ kW}}$$

(c) For the heat exchanger

$$e = (T_d - T_2)/(T_d - T_i) \quad \text{and} \quad T_1 - T_i = T_d - T_2$$

Considering the heat pump

$$COP = \frac{T_m - T_1}{(T_m - T_1) - (T_2 - T_e)}$$

Substituting the numerical values

$0.7 = (40 - T_2)/(40 - 20)$, therefore $T_2 = 26 \,°C$
$T_1 - 20 = 40 - 26$, therefore $T_1 = 34 \,°C$
$T_m - 34 = 2(T_m - 34 - 26 + 10)$, therefore $T_m = 66 \,°C$

The heat input is now $0.3(4.2)(70 - 66) = 5.0$ kW. There is also the power input to the heat pump (to drive the compressor):

$$mC_p(T_m - T_1)/COP = 0.3(4.2)(66 - 34)/2 = 20.2 \text{ kW}$$

The total energy input is therefore $5.0 + 20.2 = \mathbf{25.2 \text{ kW}}$.

The use of a heat exchanger only reduces the energy input from 63.0 to 45.4 kW, and the heat exchanger + heat pump reduces it still further to 25.2 kW. The energy saving must be balanced against the costs of an exchanger and heat pump to assess the economic viability.

10.8 Waste heat boiler

A furnace is fired with 6 kg/s of fuel oil, CV = MJ/kg. The analysis of the fuel, by mass, is 85 % C, 15 % H_2. The flue gas contains 9.35 % CO_2 (by volume) and leaves the furnace at 1050 K. C_p = 1.15 kJ/kg K.

The flue gas passes through a waste-heat boiler, leaving at 575 K. Dry steam at 12 bar is evaporated from feedwater at 290 K.

Determine the steam generation rate.

The overall coefficient = 30 W/m² K. Assuming the boiler is a single-pass (a) parallel-flow exchanger, (b) contraflow exchanger, estimate the heating area required in the boiling water, and evaporation sections.

Solution The combustion air is calculated and hence the flue-gas flow rate. Considering 1 kg oil, and expressing the combustion equation in terms of k/moles,

$$\frac{0.85}{12}[C] + \frac{0.15}{2}[H_2] + 0.21A\,[O_2] + 0.79A\,[N_2] \rightarrow$$

$$a\,[CO_2] + b\,[O_2] + c\,[N_2] + d\,[H_2O]$$

where A = amount of air supplied (k/moles).

Balancing the elements

C	$0.0708 = a$
H_2	$0.0750 = d$
O_2	$0.21A = a + b + \tfrac{1}{2}d$
N_2	$0.79A = c$

and the flue gas contains 9.35 % CO_2; therefore

$$\frac{9.35}{100} = \frac{a}{a + b + c}$$

Solving gives A = 0.795 kmol = 23.0 kg. The flue-gas flow is therefore 6 + 6(23) = 144 kg/s.

A heat balance on the waste heat boiler gives 144(1.15)(1050 − 575) = m(2784 − 71); therefore steam generated m = **29.1 kg/s**.

(a) In the boiler water section, the heat transfer is

$$Q = m(h_f - 71) = 29.1(798 - 71) = 21\,156 \text{ kW} = UA\,\Delta T_m$$

therefore

$$21\,156 = 0.03 A_1 \times \frac{(1050 - 290) - (T_1 - 461)}{\ln\,[(1050 - 290)/(T_1 - 461)]}$$

Also Q = 144(1.15)(1050 − T_1), therefore T_1 = 922 K, giving A_1 = **1179 m²**.

In the evaporator section

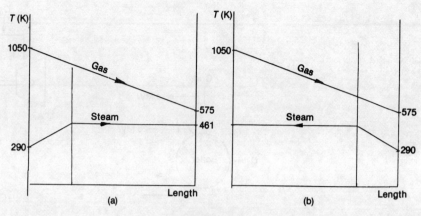

Figure 10.8

$$Q = mh_{fg} = 29.1 \times 1986 = 57\,793 \text{ kW} = 0.03 A_2 \times \frac{461 - 114}{\ln(461/114)}$$

giving $A_2 = \mathbf{7757\ m^2}$.
(b) $21\,156 = 0.03 A_1 \times (461 - 285)/\ln(461/285)$ and $A_1 = \mathbf{1927\ m^2}$
$57\,793 = 0.03 A_2 \times (589 - 461)/\ln(589/461)$ and $A_2 = \mathbf{3688\ m^2}$

10.9 CHP economics

A factory site requires 4 MW electric power and 25 MW heat for process work. The existing plant comprises a boiler, generating process steam at 3 bar, 200 °C from feedwater at 50 °C (thermal efficiency of boiler = 0.70): and the electricity is purchased at a cost of 2.5 p/kWh. The boiler is oil-fired: CV of oil = 42 MJ/kg, and the cost is £100/tonne. Calculate the energy cost/year, based on a year of 6000 h.

It is proposed to replace the plant by a gas turbine of thermal efficiency 0.25, a waste heat boiler and a steam turbine (isentropic efficiency 0.60), using steam at 20 bar, 250 °C, and exhausting to the process.

Assuming that the heat in the gas turbine exhaust gas = 0.6 of the heat input, that 0.5 of that heat is transferred in the waste heat boiler, and that surplus electric power can be exported and sold at 2.0 p/kWh, determine the energy cost/year.

The capital costs (£/kW) of the plant are: gas turbine 220, steam turbine 200, waste-heat boiler 100. Would you recommend the replacement scheme?

Solution In the existing scheme, $mh_1 = 25\,000$ where h_1 = enthalpy at 3 bar, 200 °C = 2866 kJ/kg; hence the steam generation rate = 8.72 kg/s. The heat input to the boiler is

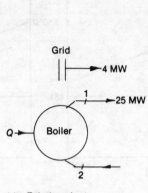
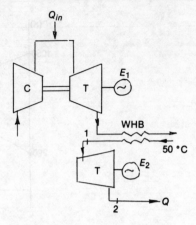

(a) Existing plant   (b) Replacement plant

Figure 10.9

$$Q_c = m(h_1 - h_2)/0.7$$
$$= 8.72(2866 - 209)/0.7 = 33\,100 \text{ kW}$$

The fuel rate = $33\,100/42\,000 = 0.79$ kg/s, therefore

annual oil cost = $0.79 \times 3600 \times 6000 \times 10^{-3} \times 100 = £1\,706\,400$
annual cost of purchased electricity = $4000 \times 6000 \times 2.5$ p
$= £600\,000$

giving a total energy cost of **£2 306 400**.

In the replacement scheme, heat transferred to steam in the waste-heat boiler is $0.5 \times 0.6 Q_{in} = 0.3 Q_{in}$ kW. For the steam turbine (using the Molier chart) $h_1 = 2904$, $h_2 = 2626$ kJ/kg and the dryness fraction at the turbine exhaust is 0.96. To meet the process requirement $mh_2 = 25\,000$; therefore $m = 9.52$ kg/s. The power generated by the steam turbine = $9.52(2904 - 2626) = 2646$ kW.

The heat absorbed by the steam in the waste-heat boiler is $9.52(2904 - 209) = 25\,656$ kW, therefore

$$Q_{in} = 25\,656/0.3 = 85\,520 \text{ kW}$$

therefore oil fuel required = $85\,520/42\,000 = 2.04$ kg/s, and

annual fuel cost = $2.04 \times 3600 \times 6000 \times 10^{-3} \times 100 = £4\,406\,400$

The power generated by the gas turbine = $0.25 \times 85\,520 = 21\,380$ kW. The total power generated is $21\,380 + 2646 = 24\,026$ kW, and the excess of $24\,026 - 4000 = 20\,026$ kW is sold (exported) for 2.0 p/kWh. There will therefore be an annual income of $2.0 \times 20\,026 \times 6000/100 = £2\,403\,120$; therefore net energy cost = **£2 003 280/year**.

The replacement scheme would save, in energy costs, £303 120/year.

The capital cost of the new plant is

£220(21 380) + 200(2646) + 100(25 656) = £7 798 400

Hence without any consideration of increased maintenance and labour costs, interest rates, etc. the simple payback time would be 7 798 400/303 120 = **25.7 years**. The replacement is therefore uneconomic.

The scheme may be economic if a cheaper fuel were available, such as coal. There are problems of corrosion due to the coal ash in a coal-fired gas turbine.

10.10 District heating

A district heating scheme uses hot water at 90 °C from a power station. The hot water is conveyed by a pipe of diameter D (m), length L (m). The fixed cost is 0.06 p/MJ at each end, and the annual capital and maintenance costs are given by

$$C = £(130D - 4 + 3 \times 10^{-7}P^3/D^5)L$$

where P = energy transmitted (MW).

Investigate the economic length of pipe, using the optimum diameter in each case, to compare with the use of a boiler on site at a unit cost of 0.15 p/MJ. The maximum demand is 10 MW.

Solution The total annual cost of the scheme is

$$C = L(130D - 4 + 3 \times 10^{-7}P^3/D^5)$$
$$+ 0.06(P) \times 3600 \times 24 \times 365/100$$
$$= L(130D - 4 + 3 \times 10^{-7}P^3/D^5) + 18\,922P$$

If the pipe diameter is an optimum, $dC/dD = 0$, therefore $0 = L(130 - 15 \times 10^{-7}P^3/D^6)$, so $D^6 = 15 \times 10^{-7}P^3/130$ and $D = 0.0475\sqrt{P}$. Substituting for D in terms of P gives

$$C = 18\,922P + L(6.175\sqrt{P} - 4 + 1.235\sqrt{P})$$
$$= 18\,922P + L(7.41\sqrt{P} - 4)$$

In the case of the on-site boiler, the annual cost is

$$C' = P \times 3600 \times 24 \times 365 \times 0.15/100 = 47\,304P$$

Table 10.6

P (MW)	1	2	3	4	5	6	7	8	9	10
C' (£)	47 304	94 068	141 102	188 136	235 170	282 204	329 238	376 272	423 333	473 040
L (m)	8 323	8 760	9 638	10 492	11 290	12 034	12 731	13 389	14 012	14 605
D (mm)	47.5	67.2	82.3	190	106	116	126	134	143	150

Problems

1 A domestic building is rectangular cross-section, 8 m × 7 m, and height 7 m. The roof area is 70 m². Calculate the heat loss from the building when the inside temperature = 15 °C, and the outside ambient temperature = 0 °C.

The overall coefficients, U (W/m² K) are: walls 2.3, glazing 5.0, roof 1.5.

Assume that the glazing occupies 10 % of the total surface area.

The walls are insulated ($U = 1.1$), the glazing replaced by double glazing ($U = 3.0$), and the roof insulated ($U = 0.5$).

Calculate the heat loss at the same temperatures.

If the costs of insulation are: walls £2/m², roof £3/m² and double glazing £80/m² would you consider the insulation economical? Assume that the ambient temperature follows the annual pattern: 0 °C for 100 days, 10 °C for 150 days, and 15 °C otherwise. The heating is supplied for 8 h/day.

Heating cost = 0.8 p/MJ.

Answer 9670 W, 4590 W: no – but could be without double glazing

2 A CHP scheme for a paper mill uses a gas turbine, waste-heat boiler and back-pressure steam turbine.

Gas turbine: oil fired. CV of oil = 42 MJ/kg. Firing rate = 1.03 kg/s. Power output = 9.7 MW.

Exhaust gases: flow rate = 62 kg/s. C_p = 1.15 kJ/kg K. Gases leave the turbine at 550 °C, and the waste-heat boiler at 300 °C. Feedwater enters at 100 °C. Steam is generated at 40 bar, 425 °C.

Steam turbine: power output = 3 MW. Back-pressure = 3 bar.

Determine the thermal efficiency of the gas turbine, the air/fuel ratio, steam-flow rate, steam-turbine isentropic efficiency, condition of the steam at the turbine exit, process heat available, and overall thermal efficiency.

Answer 0.22, 59.2 kg/kg, 6.25 kg/s, 0.79, superheated to 160 °C, 17.4 MW, 0.70

3 An industrial plant requires 3 MW of electric power and 10 kg/s of process steam at 1½ bar, 170 °C.

Investigate the attractiveness of three proposed schemes on the basis of energy cost. Assume in all cases that the boiler efficiency = 0.8, cost of coal = 0.1 p/MJ, purchased electricity = 0.7 p/MJ and surplus electricity can be exported and sold at 0.6 p/MJ.

(a) boiler, generating steam at 8 bar, 350 °C from feedwater at 20 °C, and using

   (i) a back-pressure turbine,
   (ii) a pass-out condensing turbine, exhausting at 0.10 bar.

Assume a straight condition line on the $h-s$ chart. Steam flow rate = 12 kg/s,

(b) boiler, on site, generating process steam, and electricity purchased.

Answer   costs: 3.56, 3.60, 5.96 p/s

4 A CHP scheme consists of the following:

Boiler: generating steam at 50 bar, 400 °C, thermal efficiency 0.75. Coal fired: $CV = 25$ MJ/kg.

Steam turbine: pass-out condensing at 0.05 bar, steam extracted at 5 bar. Stage isentropic efficiency = 0.8.

Process heat: latent heat extracted from the steam, the liquid going to waste. 30 MW required.

Hot-well: receives the turbine condensate and make-up water at 15 °C.

The power output = 10 MW. Determine the boiler steam and process steam-flow rates, feedwater temperature to the boiler, coal-firing rate, and overall thermal efficiency.

Answer   18.4, 14.2 kg/s; 19 °C; 3.06 kg/s; 0.52

5 A factory requires 5 MW power and 4 MW of process heat (in the form of steam at 10 bar, 180 °C).

Four schemes are proposed:

(a) Purchase power at a cost of 2.2 p/kWh. Process steam generated in a coal-fired boiler on site. Boiler efficiency = 0.75. Cost of coal = 0.1 p/MJ.

(b) Use a pass-out condensing turbine, taking steam at 30 bar, 350 °C and exhausting at 0.10 bar. Turbine isentropic efficiency (both stages) = 0.82. Boiler efficiency = 0.75, and coal-fired. Feedwater temperature = 25 °C.

(c) Use a Diesel generator and waste-heat boiler. Efficiency of Diesel engine = 0.30, and the heat in the exhaust gas = 0.3 × heat input. Cost of Diesel oil = 0.18 p/MJ. Heat transferred to steam = 0.7 × heat in exhaust gas.

(d) Use a gas turbine and waste-heat boiler. Turbine efficiency = 0.25, and the heat in the exhaust gas = 0.6 × heat input. Cost of oil fuel = 0.18 p/MJ.

The capital costs can be taken as follows: £/kW output:

steam boiler and waste-heat boiler: 100
steam turbine   : 200
gas turbine   : 220
diesel engine   : 300

Assume that the process heat is the priority, and surplus electricity can be sold at 2.5 p/kWh.

Compare the energy and capital costs of each scheme, and state which scheme you would consider to be the most attractive.

Answer Energy costs £129, 104, 106, 119/hour; capital costs (in £10⁶): 0.4, 3.2, 1.9, 0.9

6 An aero jet engine is used with a free power turbine to generate electrical power and process steam.

Ambient conditions are 1 bar, 293 K.

Engine fuel: $CV = 41$ MJ/kg, cost = 0.14 p/MJ. Compressor and turbine isentropic efficiency = 0.85 and 0.89 respectively. Pressure ratio = 10:1. Air/fuel ratio = 40 kg/kg. Maximum temperature = 1400 K.

Calculate the power generated/kg fuel and thermal efficiency of the turbine.

The exhaust gases pass through a waste-heat boiler, leaving at 500 K. Steam is generated at 10 bar, 250 °C from feedwater at 20 °C.

Determine the steam generation rate and process heat available, and the overall thermal efficiency, given the data in Table 10.7.

Table 10.7

	C_p (kJ/kg K)	γ
Air	1.005	1.40
Gases	1.15	1.33

Answer 11.3 MJ, 0.28; 5.7 kg/s, 16.3 MJ; 0.67

7 A dishwashing machine uses water at 90 °C and discharges water at 70 °C. The hot-water supply is from a calorifier in which heat is transferred from dry steam at 2 bar (leaving the calorifier coil as saturated liquid) to water entering at 10 °C. The steam rate is 1.8 kg/s, and the water flow rate 0.39 kg/s.

The thermal efficiency is to be improved by passing the waste water through a heat exchanger to preheat the cold water supply.

Analyse the possible savings in the heat input.

8 An industrial site requires an electric power load as shown in Table 10.8.

Table 10.8

L (kW)	1400	1000	500	100
h	1000	4000	1000	2760

Two diesel engine types are available (Table 10.9).

Table 10.9

Type	Installation cost £/kW	Input I (kW)
A	200	$750 + 1.5L + 1.8 \times 10^{-6} L^3$
B	230	$570 + 1.2L + 1.8 \times 10^{-6} L^3$

Assuming that no reserve capacity is required, and the fixed charge/year is 11 % of the installation cost, determine the total annual cost using each type.

Fuel cost = 0.16 p/MJ. Cost of purchased power = 2.5 p/kWh.

Which type would you consider to be the most economic?

Briefly outline how the overall plant efficiency could be improved.

Answer A £193 260, B £178 120: type B. Waste heat recovery

9 A hot-water tank is to be insulated so that the heat supply is minimized.

Water temperature = 60 °C. Ambient water temperature = 10 °C. Thermal conductivity of insulation = 0.035 W/m K. Tank dimensions: 2 m × 1 m × 1 m. Cost of heat = 0.15 p/MJ. Cost of insulation = $80x^{0.8}$ p/m², where x = thickness of insulation (cm).

Assuming continuous operation and an insulation life of 10 years, determine the minimum cost and optimum insulation thickness.

Answer £38, 13.1 cm

10 A hot-water boiler uses 9.0 kg/h of fuel, of calorific value 42 MJ/kg. The flue gases leave the boiler at 350 °C, and the specific heat of the gases can be taken as 1.15 kJ/kg K. The combustion efficiency (ratio of the heat used to heat the water to the heat released by combustion of the fuel) is given by $\eta = 1 - 1650/(m_a + 9)^2$, where m_a = air supplied (kg/h).

Determine the air/fuel ratio for maximum heat transfer to the water, and the thermal efficiency of the boiler at that condition. Ambient temperature = 15 °C.

The water enters the boiler at 20 °C and is raised to a temperature of 80 °C. Determine the water flow rate.

Answer 15.36 kg/kg, 0.77, 1750 kg/h

11 A steam turbine rejects 14 MW heat in the condenser. The cooling water is circulated through a cooling tower, of area A (m²).

Capital cost of tower = £800A
Pumping cost = £25 × 10^{-5} m³
Heat transfer in tower = $1.5 A m^{1.2}$ (W)

where m = cooling water flow rate (kg/s).

Plot a graph of cost against area and flow rate, and determine the flow rate and area at the optimum conditions (minimum cost).
Answer 1295 kg/s, 1719 m^2

12 A boiler uses fuel oil (CV = 42 MJ/kg, cost £100/tonne) to raise steam for 16 hours/day, 250 days/year. Boiler efficiency = 0.80. Firing rate = $15\frac{1}{2}$ kg/h.

It is proposed to use refuse as the fuel instead of fuel oil. CV of refuse = 18 MJ/kg, and the boiler efficiency is reduced to 0.70 by burning the refuse. The refuse available/day = 4 tonnes.

The installation cost of the conversion to refuse firing is £150 000. Investigate the economic viability of the conversion.

Briefly outline the factors that would determine the adoption of refuse incineration.
Answer 2.4 years payback period

13 Two diesel engines are used to supply 25 MW of electric power with an overall efficiency of 0.43. Each engine produces 94 000 kg/h of exhaust gas at 370 °C. The exhaust heat is to be used to generate steam for a turbo alternator to give additional power.

Estimate the increase in power output and percentage gain in thermal efficiency.

Steam supply conditions = 9 bar, 325 °C
Condenser pressure = 0.09 bar
Feedwater temperature (using engine coolant circuit) = 80 °C
Turbine isentropic efficiency = 0.8
Boiler flue gas exit temperature = 180 °C
Mean C_p for exhaust gases = 1.15 kJ/kg K.
Answer 2.7 MW; 11.6 %

14 A factory requires 720 kW power, 4320 kW process heat. The process heat is supplied by a low-pressure boiler of efficiency 0.8 and the power purchased from the CEGB power supply (generated at an efficiency of 0.27). Calculate the effective plant efficiency, defined as the power + process load/heat supplied at the factory and power station.

A new development envisages an increase in the power load to 2500 kW, the process load remaining the same. A suitable open-cycle gas turbine unit is available having the following characteristics:

maximum cycle temperature: 1000 K
pressure ratio: 6:5
compressor inlet temperature: 290 K
isentropic efficiency, compressor: 0.8
                        turbine   : 0.86

The exhaust gases from the gas turbine are to be fed into a low pressure boiler, efficiency 0.8, to provide the required process heat.

Estimate the exhaust gas exit temperature from the boiler, and the new effective plant efficiency (power + process load/heat supplied to gas turbine), given the data in Table 10.10.

Table 10.10

	C_p (kJ/kg K)	γ
Air	1.005	1.4
All gases	1.15	1.33

Answer 0.62; 466 K, 0.51

Further reading

The list of textbooks detailed is by no means complete or exhaustive, and is intended to indicate a selection which has been found useful by the author.

Boxer, G., *Applications of Engineering Thermodynamics*, Macmillan, 1979

Dixon, S.L., *Fluid Mechanics: Thermodynamics of Turbomachinery* (3rd edn), Pergamon, 1978

Eastop, T.D. and McConkey, A., *Applied Thermodynamics* (4th edn), Longman, 1986

Haywood, R.W., *Analysis of Engineering Cycles* (3rd edn), Pergamon, 1980

Kreith, F. and Black, W.Z., *Basic Heat Transfer*, Harper & Row, 1980

Rogers, G.F.C. and Mayhew, Y.R., *Engineering Thermodynamics* (3rd edn), Longman, 1980

Simonson, J.R., *Engineering Heat Transfer*, Macmillan, 1975

Index

area, critical, 27
air conditioning, 162

Biot number, 125, 130

calorific value, 1
capacity factor, 45
Carnot cycle, 33, 36
characteristics
   diesel engine, 199
   petrol engine, 62
   pump, 97, 99
   steam turbine, 45
   turbomachine, 93
coefficient
   convection, 125
   of performance, 36, 47, 157
   overall, 126
   power, 181
   thrust, 20, 27
cogeneration, 195
compression ratio, 51
compressor
   axial flow, 110
   centrifugal, 109
   reciprocating, 96, 112
convection, 125, 138
convection coefficient, 125
conduction, 124, 128
conduction transient, 135
cooling tower, 161
costs, 45, 192, 205
cut-off ratio, 51
cycles
   air standard, 51, 54
   binary vapour, 35, 42
   Carnot, 33, 36
   Joule, 72
   open, 76
   Rankine, 34, 36
   regenerative, 34, 40, 200
   reheat, 34, 38, 74, 78
   reversed Carnot, 36
   vapour compression, 47, 156, 162

dew point, 7, 159
diffuser, 97
diffusivity, 124
dissociation, 2, 8

district heating, 206
doubling time, 185
draft tube, 101

effectiveness, 11, 78, 128, 202
efficiency
   air standard, 51, 55
   diffuser, 94, 109
   fin, 127, 130
   hydraulic, 94, 97, 101
   isentropic, 73, 83, 94, 109
   isothermal, 97
   nozzle, 19, 24, 94
   propulsive, 20
   small stage, 95, 102
   solar collector, 177
   stage, 105
   volumetric, 53, 61, 64, 112
emissivity, 126
equilibrium constant, 3, 8
Euler head, 92
excess air, 2

fins, 126, 130
flame temperature, adiabatic, 3, 9
flash chamber, 156, 165
Fourier law, 124
Fourier number, 125
fuel cell, 174
fuel consumption, 184

geometrical factor, 127, 142
geothermal energy, 173, 183
Grashof number, 126

heat exchanger, 78, 128, 202
heat pump, 157, 166, 202
humidity
   relative, 155
   specific, 155

indicator diagram, 52
intercooling, 74, 79, 112

kmol, 2
Kirchhoff law, 126

Mach number, 19, 22
mean effective pressure, 51, 55, 62

nozzle, 18, 24
number of transfer units (NTU), 128
number
   Biot, 125, 130
   Fourier, 125
   Grashof, 126
   Mach, 19, 22
   Nusselt, 126, 130
   Prandtl, 126
   Reynolds, 126, 130

ocean thermal energy conversion (OTEC), 44
overall coefficient, 126

photovoltaic cell, 179
pinch point, 197
Prandtl number, 126
pressure
   partial, 3, 155
   mean effective, 51, 55
   ratio, critical, 18
pump, centrifugal, 97

radiation, 125, 140
radiosity, 142
Rankine cycle, 34, 36
reaction, degree of, 96, 107
refrigerating effect, 36
refrigeration cycle, 47
refuse incineration, 183
reheat factor, 95, 104
Reynolds number, 126, 130
rocket propulsion, 19

shape factor, 126
slip factor, 95
solar cell, 179
   collector, 175
   energy, 173
   radiation, 140
specific
   fuel consumption, 11, 19, 53, 61
   impulse, 19, 22, 27
   heat, 58, 82
   speed, 90, 99
   steam consumption, 33, 39, 41
Stefan–Boltzmann law, 127

temperature
  critical, 18
  flame, 3, 9
thermocouple, 147
thrust, 19, 22, 27
thrust coefficient, 20, 22, 27
tidal energy, 172

transient conduction, 135
transmittivity, 126, 146
turbine
  gas, 72
  hydraulic, 100
  steam, 105, 107, 192

wave energy, 173
waste heat boiler, 197, 204
wind energy, 173
windmill, 181
work done factor, 95, 110